Survey of Text Mining II

Michael W. Berry • Malu Castellanos
Editors

Survey of Text Mining II

Clustering, Classification, and Retrieval

Michael W. Berry, BS, MS, PhD
Department of Computer Science
University of Tennessee, USA

Malu Castellanos, PhD
Hewlett-Packard Laboratories
Palo Alto, California, USA

ISBN 978-1-84996-713-6   e-ISBN 978-1-84800-046-9
DOI: 10.1007/978-1-84800-046-9

British Library Cataloguing in Publication Data
A catalogue record for this book is available from the British Library

Printed on acid-free paper.

9 8 7 6 5 4 3 2 1

Springer Science+Business Media
springer.com

Preface

As we enter the third decade of the World Wide Web (WWW), the textual revolution has seen a tremendous change in the availability of online information. Finding information for just about any need has never been more automatic—just a keystroke or mouseclick away. While the digitalization and creation of textual materials continues at light speed, the ability to navigate, mine, or casually browse through documents too numerous to read (or print) lags far behind.

What approaches to text mining are available to efficiently organize, classify, label, and extract relevant information for today's information-centric users? What algorithms and software should be used to detect emerging trends from both text streams and archives? These are just a few of the important questions addressed at the Text Mining Workshop held on April 28, 2007, in Minneapolis, MN. This workshop, the fifth in a series of annual workshops on text mining, was held on the final day of the Seventh SIAM International Conference on Data Mining (April 26–28, 2007).

With close to 60 applied mathematicians and computer scientists representing universities, industrial corporations, and government laboratories, the workshop featured both invited and contributed talks on important topics such as the application of techniques of machine learning in conjunction with natural language processing, information extraction and algebraic/mathematical approaches to computational information retrieval. The workshop's program also included an Anomaly Detection/Text Mining competition. NASA Ames Research Center of Moffett Field, CA, and SAS Institute Inc. of Cary, NC, sponsored the workshop.

Most of the invited and contributed papers presented at the 2007 Text Mining Workshop have been compiled and expanded for this volume. Several others are revised papers from the first edition of the book. Collectively, they span several major topic areas in text mining:

 I. Clustering,
 II. Document retrieval and representation,
 III. Email surveillance and filtering, and
 IV. Anomaly detection.

In Part I (Clustering), Howland and Park update their work on cluster-preserving dimension reduction methods for efficient text classification. Likewise, Senellart and Blondel revisit thesaurus construction using similarity measures between vertices in graphs. Both of these chapter were part of the first edition of this book (based on a SIAM text mining workshop held in April 2002). The next three chapters are completely new contributions. Zeimpekis and Gallopoulos implement and evaluate several clustering schemes that combine partitioning and hierarchical algorithms. Kogan, Nicholas, and Wiacek look at the hybrid clustering of large, high-dimensional data. AlSumait and Domeniconi round out this topic area with an examination of local semantic kernels for the clustering of text documents.

In Part II (Document Retrieval and Representation), Kobayashi and Aono revise their first edition chapter on the importance of detecting and interpreting minor document clusters using a vector space model based on principal component analysis (PCA) rather than the popular latent semantic indexing (LSI) method. This is followed by Xia, Xing, Qi, and Li's chapter on applications of semidefinite programming in XML document classification.

In Part III (Email Surveillance and Filtering), Bader, Berry, and Browne take advantage of the Enron email dataset to look at topic detection over time using PARAFAC and multilinear algebra. Gansterer, Janecek, and Neumayer examine the use of latent semantic indexing to combat email spam.

In Part IV (Anomaly Detection), researchers from the NASA Ames Research Center share approaches to anomaly detection. These techniques were actually entries in a competition. held as part of the workshop. The top three finishers in the competition were: Cyril Goutte of NRC Canada, Edward G. Allan, Michael R. Horvath, Christopher V. Kopek, Brian T. Lamb, and Thomas S. Whaples of Wake Forest University (Michael W. Berry of the University of Tennessee was their advisor), and an international group from the Middle East led by Mostafa Keikha. Each chapter provides an explanation of its approach to the contest.

This volume details the state-of-the-art algorithms and software for text mining from both the academic and industrial perspectives. Familiarity or coursework (undergraduate-level) in vector calculus and linear algebra is needed for several of the chapters. While many open research questions still remain, this collection serves as an important benchmark in the development of both current and future approaches to mining textual information.

Acknowledgments

The editors would like to thank Murray Browne of the University of Tennessee and Catherine Brett of Springer UK in coordinating the management of manuscripts among the authors, editors, and the publisher.

Michael W. Berry and Malu Castellanos
Knoxville, TN and Palo Alto, CA
August 2007

Contents

Contributors

Edward G. Allan
> Department of Computer Science
> Wake Forest University
> P.O. Box 7311
> Winston-Salem, NC 27109
> Email: allaeg3@wfu.edu

Loulwah Alsumait
> Department of Computer Science
> George Mason University
> 4400 University Drive MSN 4A4
> Fairfax, VA 22030
> Email: lalsumai@gmu.edu

Masaki Aono
> Department of Information and Computer Sciences, C-511
> Toyohashi University of Technology
> 1-1 Hibarigaoka, Tempaku-cho
> Toyohashi-shi, Aichi 441-8580
> Japan
> Email: aono@ics.tut.ac.jp

Brett W. Bader
> Sandia National Laboratories
> Applied Computational Methods Department
> P.O. Box 5800
> Albuquerque, NM 87185-1318
> Email: bwbader@sandia.gov
> Homepage: http://www.cs.sandia.gov/~bwbader

Michael W. Berry
> Department of Electrical Engineering and Computer Science
> University of Tennessee
> 203 Claxton Complex
> Knoxville, TN 37996-3450
> Email: berry@eecs.utk.edu
> Homepage: http://www.cs.utk.edu/~berry

Vincent D. Blondel
    Division of Applied Mathematics
    Université de Louvain
    4, Avenue Georges Lemaître
    B-1348 Louvain-la-neuve
    Belgium
    Email: blondel@inma.ucl.ac.be
    Homepage: http://www.inma.ucl.ac.be/~blondel

Murray Browne
    Department of Electrical Engineering and Computer Science
    University of Tennessee
    203 Claxton Complex
    Knoxville, TN 37996-3450
    Email: mbrowne@eecs.utk.edu

Malú Castellanos
    IETL Department
    Hewlett-Packard Laboratories
    1501 Page Mill Road MS-1148
    Palo Alto, CA 94304
    Email: malu.castellanos@hp.com

Pat Castle
    Intelligent Systems Division
    NASA Ames Research Center
    Moffett Field, CA 94035
    Email: pcastle@email.arc.nasa.gov

Santanu Das
    Intelligent Systems Division
    NASA Ames Research Center
    Moffett Field, CA 94035
    Email: sdas@email.arc.nasa.gov

Carlotta Domeniconi
    Department of Computer Science
    George Mason University
    4400 University Drive MSN 4A4
    Fairfax, VA 22030
    Email: carlotta@ise.gmu.edu
    Homepage: http://www.ise.gmu.edu/~carlotta

Efstratios Gallopoulos
   Department of Computer Engineering and Informatics
   University of Patras
   26500 Patras
   Greece
   Email: stratis@hpclab.ceid.upatras.gr
   Homepage: http://scgroup.hpclab.ceid.upatras.gr/faculty/stratis/stratise.html

Wilfried N. Gansterer
   Research Lab for Computational Technologies and Applications
   University of Vienna
   Lenaugasse 2/8
   A - 1080 Vienna
   Austria
   Email: wilfried.gansterer@univie.ac.at

Cyril Goutte
   Interactive Language Technologies
   NRC Institute for Information Technology
   283 Boulevard Alexandre Taché
   Gatineau, QC J8X 3X7
   Canada
   Email: cyril.goutte@nrc-cnrc.gc.ca
   Homepage: http://iit-iti.nrc-cnrc.gc.ca/personnel/goutte_cyril_f.html

Michael R. Horvath
   Department of Computer Science
   Wake Forest University
   P.O. Box 7311
   Winston-Salem, NC 27109
   Email: horvmr5@wfu.edu

Peg Howland
   Department of Mathematics and Statistics
   Utah State University
   3900 Old Main Hill
   Logan, UT 84322-3900
   Email: peg.howland@usu.edu
   Homepage: http://www.math.usu.edu/~howland

Andreas G. K. Janecek
Research Lab for Computational Technologies and Applications
University of Vienna
Lenaugasse 2/8
A - 1080 Vienna
Austria
Email: andreas.janecek@univie.ac.at

Mostafa Keikha
Department of Electrical and Computer Engineering
University of Tehran
P.O. Box 14395-515, Tehran
Iran
Email: m.keikha@ece.ut.ac.ir

Mei Kobayashi
IBM Research, Tokyo Research Laboratory
1623-14 Shimotsuruma, Yamato-shi
Kanagawa-ken 242-8502
Japan
Email: mei@jp.ibm.com
Homepage: http://www.trl.ibm.com/people/meik

Jacob Kogan
Department of Mathematics and Statistics
University of Maryland, Baltimore County
1000 Hilltop Circle
Baltimore, MD 21250
Email: kogan@math.umbc.edu
Homepage: http://www.math.umbc.edu/~kogan

Christopher V. Kopek
Department of Computer Science
Wake Forest University
P.O. Box 7311
Winston-Salem, NC 27109
Email: kopecv5@wfu.edu

Brian T. Lamb
Department of Computer Science
Wake Forest University
P.O. Box 7311
Winston-Salem, NC 27109
Email: lambbt5@wfu.edu

Qi Li
    Department of Computer Science
    Western Kentucky University
    1906 College Heights Boulevard #11076
    Bowling Green, KY 42101-1076
    Email: qi.li@wku.edu
    Homepage: http://www.wku.edu/~qi.li

Robert Neumayer
    Institute of Software Technology and Interactive Systems
    Vienna University of Technology
    Favoritenstraße 9-11/188/2
    A - 1040 Vienna
    Austria
    Email: neumayer@ifs.tuwien.ac.at

Charles Nicholas
    Department of Computer Science and Electrical Engineering
    University of Maryland, Baltimore County
    1000 Hilltop Circle
    Baltimore, MD 21250
    Email: nicholas@cs.umbc.edu
    Homepage: http://www.cs.umbc.edu/~nicholas

Farhad Oroumchian
    College of Information Technology
    University of Wollongong in Dubai
    P.O. Box 20183, Dubai
    U.A.E.
    Email: farhadoroumchian@uowdubai.ac.ae

Matthew E. Otey
    Intelligent Systems Division
    NASA Ames Research Center
    Moffett Field, CA 94035
    Email: otey@email.arc.nasa.gov

Haesun Park
    Division of Computational Science and Engineering
    College of Computing
    Georgia Institute of Technology
    266 Ferst Drive
    Atlanta, GA 30332-0280
    Email: hpark@cc.gatech.edu
    Homepage: http://www.cc.gatech.edu/~hpark

Houduo Qi
Department of Mathematics
University of Southampton, Highfield
Southampton SO17 1BJ, UK
Email: hdqi@soton.ac.uk
Homepage: http://www.personal.soton.ac.uk/hdqi

Narjes Sharif Razavian
Department of Electrical and Computer Engineering
University of Tehran
P.O. Box 14395-515, Tehran
Iran
Email: n.razavian@ece.ut.ac.ir

Hassan Seyed Razi
Department of Electrical and Computer Engineering
University of Tehran
P.O. Box 14395-515, Tehran
Iran
Email: seyedraz@ece.ut.ac.ir

Pierre Senellart
INRIA Futurs & Université Paris-Sud
4 rue Jacques Monod
91893 Orsay Cedex
France
Email: pierre@senellart.com
Homepage: http://pierre.senellart.com/

Ashok N. Srivastava
Intelligent Systems Division
NASA Ames Research Center
Moffett Field, CA 94035
Email: ashok@email.arc.nasa.gov

Thomas S. Whaples
Department of Computer Science
Wake Forest University
P.O. Box 7311
Winston-Salem, NC 27109
Email: whapts3@wfu.edu

Mike Wiacek
Google, Inc.
1600 Amphitheatre Parkway
Mountain View, CA 94043
Email: mjwiacek@google.com

Zhonghang Xia
    Department of Computer Science
    Western Kentucky University
    1906 College Heights Boulevard #11076
    Bowling Green, KY 42101-1076
    Email: zhonghang.xia@wku.edu
    Homepage: http://www.wku.edu/~zhonghang.xia

Guangming Xing
    Department of Computer Science
    Western Kentucky University
    1906 College Heights Boulevard #11076
    Bowling Green, KY 42101-1076
    Email: guangming.xing@wku.edu
    Homepage: http://www.wku.edu/~guangming.xing

Dimitrios Zeimpekis
    Department of Computer Engineering and Informatics
    University of Patras
    26500 Patras
    Greece
    Email: dsz@hpclab.ceid.upatras.gr

Part I

Clustering

1

Cluster-Preserving Dimension Reduction Methods for Document Classification

Peg Howland and Haesun Park

Overview

In today's vector space information retrieval systems, dimension reduction is imperative for efficiently manipulating the massive quantity of data. To be useful, this lower dimensional representation must be a good approximation of the original document set given in its full space. Toward that end, we present mathematical models, based on optimization and a general matrix rank reduction formula, which incorporate a priori knowledge of the existing structure. From these models, we develop new methods for dimension reduction that can be applied regardless of the relative dimensions of the term-document matrix. We illustrate the effectiveness of each method with document classification results from the reduced representation. After establishing relationships among the solutions obtained by the various methods, we conclude with a discussion of their relative accuracy and complexity.

1.1 Introduction

The vector space information retrieval system, originated by Gerard Salton [Sal71, SM83], represents documents as vectors in a vector space. The document set comprises an $m \times n$ term-document matrix, in which each column represents a document, and the (i, j)th entry represents a weighted frequency of term i in document j. Since the data dimension m may be huge, a lower dimensional representation is imperative for efficient manipulation.

Dimension reduction is commonly based on rank reduction by the truncated singular value decomposition (SVD). For any matrix $A \in \mathbb{R}^{m \times n}$, its SVD can be defined as

$$A = U \Sigma V^T, \tag{1.1}$$

where $U \in \mathbb{R}^{m \times m}$ and $V \in \mathbb{R}^{n \times n}$ are orthogonal, $\Sigma = diag(\sigma_1 \cdots \sigma_p) \in \mathbb{R}^{m \times n}$ with $p = \min(m, n)$, and the singular values are ordered as $\sigma_1 \geq \sigma_2 \geq \cdots \sigma_p \geq 0$ [GV96, Bjö96]. Denoting the columns of U, or left singular vectors, by u_i, and the columns of V, or right singular vectors, by v_i, and the rank of A by q, we write

$$A = \sum_{i=1}^{q} \sigma_i u_i v_i^T. \qquad (1.2)$$

For $l < q$, the truncated SVD

$$A \approx \sum_{i=1}^{l} \sigma_i u_i v_i^T$$

provides the rank-l approximation that is closest to the data matrix in L_2 norm or Frobenius norm[GV96]. This is the main tool in principal component analysis (PCA) [DHS01], as well as in latent semantic indexing (LSI) [DDF+90, BDO95] of documents.

If the data form clusters in the full dimension, the goal may change from finding the best lower dimensional representation to finding the lower dimensional representation that best preserves this cluster structure. That is, even after dimension reduction, a new document vector should be classified with the appropriate cluster. Assuming that the columns of A are grouped into clusters, rather than treating each column equally regardless of its membership in a specific cluster, as is done with the SVD, the dimension reduction methods we will discuss attempt to preserve this information. This is important in information retrieval, since the reduced representation itself will be used extensively in further processing of data.

In applied statistics/psychometrics [Har67, Hor65], techniques have been developed to factor an attribute-entity matrix in an analogous way. As argued in [HMH00], the components of the factorization are important, and "not just as a mechanism for solving another problem." This is one reason why they are well suited for the problem of finding a lower dimensional representation of text data. Another reason is their simplicity—having been developed at a time when additions and subtractions were significantly cheaper computations than multiplications and divisions—their creators used sign and binary vectors extensively. With the advent of modern computers, such methods have become overshadowed by more accurate and costly algorithms in factor analysis [Fuk90]. Ironically, modern applications often have to handle very high-dimensional data, so the accuracy of the factors can sometimes be compromised in favor of algorithmic simplicity.

In this chapter, we present dimension reduction methods derived from two perspectives. The first, a general matrix rank reduction formula, is introduced in Section 1.2. The second, linear discriminant analysis (LDA), is formulated as trace optimization, and extended in Section 1.3 using the generalized singular value decomposition (GSVD). To reduce the cost of the LDA/GSVD algorithm, we incorporate it as the second stage after PCA or LSI. We establish mathematical equivalence in Section 1.4 by expressing both PCA and LSI in terms of trace optimization. Finally, Section 1.5 combines the two perspectives by making a factor analysis approximation in the first stage.

1.2 Dimension Reduction in the Vector Space Model (VSM)

Given a term-document matrix

$$A = [a_1 \quad a_2 \quad \cdots \quad a_n] \in \mathbb{R}^{m \times n},$$

we want to find a transformation that maps each document vector a_j in the m-dimensional space to a vector y_j in the l-dimensional space for some $l \ll m$:

$$a_j \in \mathbb{R}^{m \times 1} \to y_j \in \mathbb{R}^{l \times 1}, \quad 1 \le i \le n.$$

The approach we discuss in Section 1.3 computes the transformation directly from A. Rather than looking for the mapping that achieves this explicitly, another approach rephrases dimension reduction as an approximation problem where the given matrix A is decomposed into two matrices B and Y as

$$A \approx BY \tag{1.3}$$

where both $B \in \mathbb{R}^{m \times l}$ with $\mathrm{rank}(B) = l$ and $Y \in \mathbb{R}^{l \times n}$ with $\mathrm{rank}(Y) = l$ are to be found. This lower rank approximation is not unique since for any nonsingular matrix $Z \in \mathbb{R}^{l \times l}$,

$$A \approx BY = (BZ)(Z^{-1}Y),$$

where $\mathrm{rank}(BZ) = l$ and $\mathrm{rank}(Z^{-1}Y) = l$.

The mathematical framework for obtaining such matrix factorizations is the Wedderburn rank reduction formula [Wed34]: If $x \in \mathbb{R}^n$ and $y \in \mathbb{R}^m$ are such that $\omega = y^T A x \ne 0$, then

$$E = A - \omega^{-1}(Ax)(y^T A) \tag{1.4}$$

has $\mathrm{rank}(E) = \mathrm{rank}(A) - 1$. This formula has been studied extensively in both the numerical linear algebra (NLA) [CF79, CFG95] and applied statistics/psychometrics (AS/P) [Gut57, HMH00] communities. In a 1995 *SIAM Review* paper [CFG95], Chu, Funderlic, and Golub show that for $x = v_1$ and $y = u_1$ from the SVD in Eq. (1.2),

$$E = A - (u_1^T A v_1)^{-1}(A v_1)(u_1^T A) = A - \sigma_1 u_1 v_1^T.$$

If repeated $q = \mathrm{rank}(A)$ times using the leading q singular vectors, this formula generates the SVD of A.

In general, starting with $A_1 = A$, and choosing x_k and y_k such that $\omega_k = y_k^T A_k x_k \ne 0$, the Wedderburn formula generates the sequence

$$A_{k+1} = A_k - \omega_k^{-1}(A_k x_k)(y_k^T A_k).$$

Adding up all the rank one updates, factoring into matrix outer product form, and truncating gives an approximation $A \approx BY$. The question becomes: what are good choices for x_k and y_k?

One answer was provided by Thurstone [Thu35] in the 1930s, when he applied the centroid method to psychometric data. To obtain an approximation of A as BY, the method uses the rank one reduction formula to solve for one column of B and one row of Y at a time. It approximates the SVD while restricting the pre- and post-factors in the rank reduction formula to sign vectors. In Section 1.5, we incorporate this SVD approximation into a two-stage process so that knowledge of the clusters from the full dimension is reflected in the dimension reduction.

1.3 Linear Discriminant Analysis and Its Extension for Text Data

The goal of linear discriminant analysis (LDA) is to combine features of the original data in a way that most effectively discriminates between classes. With an appropriate extension, it can be applied to the goal of reducing the dimension of a term-document matrix in a way that most effectively preserves its cluster structure. That is, we want to find a linear transformation G whose transpose maps each document vector a in the m-dimensional space to a vector y in the l-dimensional space ($l \ll m$):

$$G^T : a \in \mathbb{R}^{m \times 1} \rightarrow y \in \mathbb{R}^{l \times 1}.$$

Assuming that the given data are already clustered, we seek a transformation that optimally preserves this cluster structure in the reduced dimensional space.

For simplicity of discussion, we will assume that data vectors $a_1, \ldots, a_n$ form columns of a matrix $A \in \mathbb{R}^{m \times n}$, and are grouped into k clusters as

$$A = [A_1, A_2, \cdots, A_k], A_i \in \mathbb{R}^{m \times n_i}, \sum_{i=1}^{k} n_i = n. \tag{1.5}$$

Let N_i denote the set of column indices that belong to cluster i. The centroid $c^{(i)}$ is computed by taking the average of the columns in cluster i; i.e.,

$$c^{(i)} = \frac{1}{n_i} \sum_{j \in N_i} a_j$$

and the global centroid c is defined as

$$c = \frac{1}{n} \sum_{j=1}^{n} a_j.$$

Then the within-cluster, between-cluster, and mixture scatter matrices are defined [Fuk90, TK99] as

$$S_w = \sum_{i=1}^{k} \sum_{j \in N_i} (a_j - c^{(i)})(a_j - c^{(i)})^T,$$

$$S_b = \sum_{i=1}^{k} \sum_{j \in N_i} (c^{(i)} - c)(c^{(i)} - c)^T$$

$$= \sum_{i=1}^{k} n_i (c^{(i)} - c)(c^{(i)} - c)^T, \text{ and}$$

$$S_m = \sum_{j=1}^{n} (a_j - c)(a_j - c)^T,$$

respectively. The scatter matrices have the relationship [JD88]

$$S_m = S_w + S_b. \tag{1.6}$$

Applying G^T to the matrix A transforms the scatter matrices S_w, S_b, and S_m to the $l \times l$ matrices

$$G^T S_w G, \quad G^T S_b G, \quad \text{and} \quad G^T S_m G,$$

respectively.

There are several measures of cluster quality that involve the three scatter matrices [Fuk90, TK99]. When cluster quality is high, each cluster is tightly grouped, but well separated from the other clusters. Since

$$\text{trace}(S_w) = \sum_{i=1}^{k} \sum_{j \in N_i} (a_j - c^{(i)})^T (a_j - c^{(i)})$$

$$= \sum_{i=1}^{k} \sum_{j \in N_i} \| a_j - c^{(i)} \|_2^2$$

measures the closeness of the columns within the clusters, and

$$\text{trace}(S_b) = \sum_{i=1}^{k} \sum_{j \in N_i} (c^{(i)} - c)^T (c^{(i)} - c)$$

$$= \sum_{i=1}^{k} \sum_{j \in N_i} \| c^{(i)} - c \|_2^2$$

measures the separation between clusters, an optimal transformation that preserves the given cluster structure would be

$$\max_{G} \text{trace}(G^T S_b G) \quad \text{and} \quad \min_{G} \text{trace}(G^T S_w G). \tag{1.7}$$

Assuming the matrix S_w is nonsingular, classical LDA approximates this simultaneous trace optimization by finding a transformation G that maximizes

$$J_1(G) = \text{trace}((G^T S_w G)^{-1} G^T S_b G). \tag{1.8}$$

It is well-known that the J_1 criterion in Eq. (1.8) is maximized when the columns of G are the l eigenvectors of $S_w^{-1} S_b$ corresponding to the l largest eigenvalues [Fuk90]. In other words, LDA solves

$$S_w^{-1} S_b x_i = \lambda_i x_i \tag{1.9}$$

for the x_i's corresponding to the largest λ_i's. For these l eigenvectors, the maximum achieved is $J_1(G) = \lambda_1 + \cdots + \lambda_l$. Since $\text{rank}(S_b)$ of the eigenvalues of $S_w^{-1} S_b$ are greater than zero, if $l \geq \text{rank}(S_b)$, this optimal G preserves $\text{trace}(S_w^{-1} S_b)$ exactly upon dimension reduction.

For the case when S_w is singular, [HJP03] assumes the cluster structure given in Eq. (1.5), and defines the $m \times n$ matrices

$$H_w = [A_1 - c^{(1)} e^{(1)^T}, A_2 - c^{(2)} e^{(2)^T}, \ldots, A_k - c^{(k)} e^{(k)^T}] \tag{1.10}$$

$$H_b = [(c^{(1)} - c) e^{(1)^T}, (c^{(2)} - c) e^{(2)^T}, \ldots, (c^{(k)} - c) e^{(k)^T}]$$

$$H_m = [a_1 - c, \ldots, a_n - c] = A - ce^T, \tag{1.11}$$

where $e^{(i)} = (1, \ldots, 1)^T \in \mathbb{R}^{n_i \times 1}$ and $e = (1, \cdots, 1)^T \in \mathbb{R}^{n \times 1}$. Then the scatter matrices can be expressed as

$$S_w = H_w H_w^T, \quad S_b = H_b H_b^T, \quad \text{and} \quad S_m = H_m H_m^T. \tag{1.12}$$

Another way to define H_b that satisfies Eq. (1.12) is

$$H_b = [\sqrt{n_1}(c^{(1)} - c), \sqrt{n_2}(c^{(2)} - c), \ldots, \sqrt{n_k}(c^{(k)} - c)] \tag{1.13}$$

and using this $m \times k$ form reduces the storage requirements and computational complexity of the LDA/GSVD algorithm.

As the product of an $m \times n$ matrix and an $n \times m$ matrix, S_w is singular when $m > n$ [Ort87]. This means that J_1 cannot be applied when the number of available data points is smaller than the dimension of the data. In other words, classical LDA fails when the number of terms in the document collection is larger than the total number of documents (i.e., $m > n$ in the term-document matrix A). To circumvent this restriction, we express λ_i as α_i^2 / β_i^2, and the eigenvalue problem in Eq. (1.9) becomes

$$\beta_i^2 H_b H_b^T x_i = \alpha_i^2 H_w H_w^T x_i. \tag{1.14}$$

This has the form of a problem that can be solved using the GSVD of the matrix pair (H_b^T, H_w^T), as described in Section 1.3.1.

1.3.1 Generalized Singular Value Decomposition

After the GSVD was originally defined by Van Loan [Loa76], Paige and Saunders [PS81] defined the GSVD for any two matrices with the same number of columns, which we restate as follows.

Theorem 1.3.1 *Suppose two matrices $H_b^T \in \mathbb{R}^{k \times m}$ and $H_w^T \in \mathbb{R}^{n \times m}$ are given. Then for*

$$K = \begin{pmatrix} H_b^T \\ H_w^T \end{pmatrix} \quad and \quad t = rank(K),$$

there exist orthogonal matrices $U \in \mathbb{R}^{k \times k}$, $V \in \mathbb{R}^{n \times n}$, $W \in \mathbb{R}^{t \times t}$, and $Q \in \mathbb{R}^{m \times m}$ such that

$$U^T H_b^T Q = \Sigma_b (\underbrace{W^T R}_{t}, \underbrace{0}_{m-t})$$

and

$$V^T H_w^T Q = \Sigma_w (\underbrace{W^T R}_{t}, \underbrace{0}_{m-t}),$$

where

$$\Sigma_b = \underset{k \times t}{\begin{pmatrix} I_b & & \\ & D_b & \\ & & O_b \end{pmatrix}}, \quad \Sigma_w = \underset{n \times t}{\begin{pmatrix} O_w & & \\ & D_w & \\ & & I_w \end{pmatrix}},$$

and $R \in \mathbb{R}^{t \times t}$ is nonsingular with its singular values equal to the nonzero singular values of K. The matrices

$$I_b \in \mathbb{R}^{r \times r} \quad and \quad I_w \in \mathbb{R}^{(t-r-s) \times (t-r-s)}$$

are identity matrices, where

$$r = t - rank(H_w^T) \quad and \quad s = rank(H_b^T) + rank(H_w^T) - t,$$

$$O_b \in \mathbb{R}^{(k-r-s) \times (t-r-s)} \quad and \quad O_w \in \mathbb{R}^{(n-t+r) \times r}$$

are zero matrices with possibly no rows or no columns, and

$$D_b = diag(\alpha_{r+1}, \ldots, \alpha_{r+s})$$

and

$$D_w = diag(\beta_{r+1}, \ldots, \beta_{r+s})$$

satisfy

$$1 > \alpha_{r+1} \geq \cdots \geq \alpha_{r+s} > 0, \quad 0 < \beta_{r+1} \leq \cdots \leq \beta_{r+s} < 1, \qquad (1.15)$$

and $\alpha_i^2 + \beta_i^2 = 1$ for $i = r+1, \ldots, r+s$.

This form of GSVD is related to that of Van Loan [Loa76] as

$$U^T H_b^T X = (\Sigma_b, 0) \quad and \quad V^T H_w^T X = (\Sigma_w, 0), \qquad (1.16)$$

where

$$\underset{m \times m}{X} = Q \begin{pmatrix} R^{-1}W & 0 \\ 0 & I_{m-t} \end{pmatrix}.$$

This implies that

$$X^T H_b H_b^T X = \begin{pmatrix} \Sigma_b^T \Sigma_b & 0 \\ 0 & 0 \end{pmatrix}$$

and

$$X^T H_w H_w^T X = \begin{pmatrix} \Sigma_w^T \Sigma_w & 0 \\ 0 & 0 \end{pmatrix}.$$

Letting x_i represent the ith column of X, and defining

$$\alpha_i = 1, \ \beta_i = 0 \text{ for } i = 1, \ldots, r$$

and

$$\alpha_i = 0, \ \beta_i = 1 \text{ for } i = r + s + 1, \ldots, t,$$

we see that Eq. (1.14) is satisfied for $1 \leq i \leq t$. Since

$$H_b H_b^T x_i = 0 \quad \text{and} \quad H_w H_w^T x_i = 0$$

for the remaining $m-t$ columns of X, Eq. (1.14) is satisfied for arbitrary values of α_i and β_i when $t + 1 \leq i \leq m$. The columns of X are the generalized singular vectors for the matrix pair (H_b^T, H_w^T). They correspond to the generalized singular values, or the α_i/β_i quotients, as follows. The first r columns correspond to infinite values, and the next s columns correspond to finite and nonzero values. The following $t - r - s$ columns correspond to zero values, and the last $m - t$ columns correspond to the arbitrary values. This correspondence between generalized singular vectors and values is illustrated in Figure 1.1(a).

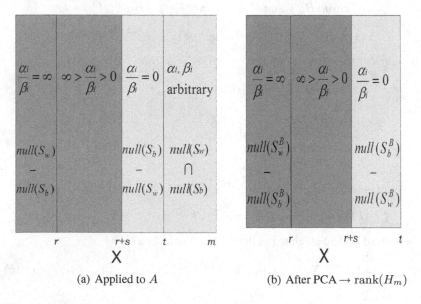

(a) Applied to A (b) After PCA $\rightarrow$ rank(H_m)

Fig. 1.1. Generalized singular vectors and their corresponding generalized singular values.

1.3.2 Application of the GSVD to Dimension Reduction

A question that remains is which columns of X to include in the solution G. If S_w is nonsingular, both $r = 0$ and $m - t = 0$, so $s = \mathrm{rank}(H_b^T)$ generalized singular values are finite and nonzero, and the rest are zero. The generalized singular vectors are eigenvectors of $S_w^{-1}S_b$, so we choose the x_i's that correspond to the largest λ_i's, where $\lambda_i = \alpha_i^2/\beta_i^2$. When the GSVD construction orders the singular value pairs as in Eq. (1.15), the generalized singular values, or the α_i/β_i quotients, are in non-increasing order. Therefore, the first s columns of X are all we need.

When $m > n$, the scatter matrix S_w is singular. Hence, the eigenvectors of $S_w^{-1}S_b$ are undefined, and classical discriminant analysis fails. [HJP03] argues in terms of the simultaneous optimization Eq. (1.7) that criterion J_1 is approximating. Letting g_j represent a column of G, we write

$$\mathrm{trace}(G^T S_b G) = \sum g_j^T S_b g_j$$

and

$$\mathrm{trace}(G^T S_w G) = \sum g_j^T S_w g_j.$$

If x_i is one of the leftmost r vectors, then $x_i \in \mathrm{null}(S_w) - \mathrm{null}(S_b)$. Because $x_i^T S_b x_i > 0$ and $x_i^T S_w x_i = 0$, including this vector in G increases the trace we want to maximize while leaving the trace we want to minimize unchanged. On the other hand, for the rightmost $m - t$ vectors, $x_i \in \mathrm{null}(S_w) \cap \mathrm{null}(S_b)$. Adding the column x_i to G has no effect on these traces, since $x_i^T S_w x_i = 0$ and $x_i^T S_b x_i = 0$, and therefore does not contribute to either maximization or minimization in Eq. (1.7). We conclude that, whether S_w is singular or nonsingular, G should be comprised of the leftmost $r + s = \mathrm{rank}(H_b^T)$ columns of X, which are shaded in Figure 1.1(a).

As a practical matter, the LDA/GSVD algorithm includes the first $k - 1$ columns of X in G. This is due to the fact that $\mathrm{rank}(H_b) \le k - 1$, which is clear from the definition of H_b given in Eq. (1.13). If $\mathrm{rank}(H_b) < k - 1$, including extra columns in G (some of which correspond to the $t - r - s$ zero generalized singular values and, possibly, some of which correspond to the arbitrary generalized singular values) will have approximately no effect on cluster preservation. As summarized in Algorithm 1.3.1, we first compute the matrices H_b and H_w from the data matrix A. We then solve for a very limited portion of the GSVD of the matrix pair (H_b^T, H_w^T). This solution is accomplished by following the construction in the proof of Theorem 1.3.1 [PS81]. The major steps are limited to the complete orthogonal decomposition [GV96, LH95] of

$$K = \begin{pmatrix} H_b^T \\ H_w^T \end{pmatrix},$$

which produces orthogonal matrices P and Q and a nonsingular matrix R, followed by the singular value decomposition of a leading principal submatrix of P, whose size is much smaller than that of the data matrix. (This $k \times t$ submatrix is specified in Algorithm 1.3.1 using the colon notation of MATLAB[1].) Finally, we assign the leftmost $k - 1$ generalized singular vectors to G.

---

[1] http://www.mathworks.com

Algorithm 1.3.1 LDA/GSVD

Given a data matrix $A \in \mathbb{R}^{m \times n}$ with k clusters and an input vector $a \in \mathbb{R}^{m \times 1}$, compute the matrix $G \in \mathbb{R}^{m \times (k-1)}$ that preserves the cluster structure in the reduced dimensional space, using

$$J_1(G) = \text{trace}((G^T S_w G)^{-1} G^T S_b G).$$

Also compute the $k-1$ dimensional representation y of a.

1. Compute H_b and H_w from A according to

$$H_b = (\sqrt{n_1}(c^{(1)} - c), \sqrt{n_2}(c^{(2)} - c), \dots, \sqrt{n_k}(c^{(k)} - c))$$

   and Eq. (1.10), respectively. (Using this equivalent but $m \times k$ form of H_b reduces complexity.)
2. Compute the complete orthogonal decomposition

$$P^T K Q = \begin{pmatrix} R & 0 \\ 0 & 0 \end{pmatrix}, \text{ where } K = \begin{pmatrix} H_b^T \\ H_w^T \end{pmatrix} \in \mathbb{R}^{(k+n) \times m}$$

3. Let $t = \text{rank}(K)$.
4. Compute W from the SVD of $P(1:k, 1:t)$, which is

$$U^T P(1:k, 1:t) W = \Sigma_A.$$

5. Compute the first $k-1$ columns of $X = Q \begin{pmatrix} R^{-1} W & 0 \\ 0 & I \end{pmatrix}$, and assign them to G.
6. $y = G^T a$

1.4 Equivalent Two-Stage Methods

Another way to apply LDA to the data matrix $A \in \mathbb{R}^{m \times n}$ with $m > n$ (and hence S_w singular) is to perform dimension reduction in two stages. The LDA stage is preceded by a stage in which the cluster structure is ignored. A common approach [Tor01, SW96, BHK97] for the first part of this process is rank reduction by the truncated singular value decomposition (SVD). A drawback of these two-stage approaches is that experimentation has been needed to determine which intermediate reduced dimension produces optimal results after the second stage.

Moreover, since either PCA or LSI ignores the cluster structure in the first stage, theoretical justification for such two-stage approaches is needed. Yang and Yang [YY03] supply theoretical justification for PCA plus LDA, for a single discriminant vector. In this section, we justify the two-stage approach that uses either LSI or PCA, followed by LDA. We do this by establishing the equivalence of the single-stage LDA/GSVD to the two-stage method, provided that the intermediate dimension after the first stage falls within a specific range. In this range S_w remains singular, and hence LDA/GSVD is required for the second stage. We also present a computationally simpler choice for the first stage, which uses QR decomposition (QRD) rather than the SVD.

1.4.1 Rank Reduction Based on the Truncated SVD

PCA and LSI differ only in that PCA centers the data by subtracting the global centroid from each column of A. In this section, we express both methods in terms of the maximization of $J_2(G) = \text{trace}(G^T S_m G)$.

If we let $G \in \mathbb{R}^{m \times l}$ be any matrix with full column rank, then essentially $J_2(G)$ has no upper bound and maximization is meaningless. Now, let us restrict the solution to the case when G has orthonormal columns. Then there exists $G' \in \mathbb{R}^{m \times (m-l)}$ such that (G, G') is an orthogonal matrix. In addition, since S_m is positive semidefinite, we have

$$\text{trace}(G^T S_m G) \leq \text{trace}(G^T S_m G) + \text{trace}((G')^T S_m G')$$
$$= \text{trace}(S_m).$$

Reserving the notation in Eq. (1.1) for the SVD of A, let the SVD of H_m be given by

$$H_m = A - ce^T = \tilde{U} \tilde{\Sigma} \tilde{V}^T. \tag{1.17}$$

Then

$$S_m = H_m H_m^T = \tilde{U} \tilde{\Sigma} \tilde{\Sigma}^T \tilde{U}^T.$$

Hence the columns of $\tilde{U}$ form an orthonormal set of eigenvectors of S_m corresponding to the non-increasing eigenvalues on the diagonal of $\Lambda = \tilde{\Sigma} \tilde{\Sigma}^T = \text{diag}(\tilde{\sigma}_1^2, \ldots, \tilde{\sigma}_n^2, 0, \ldots, 0)$. For $p = \text{rank}(H_m)$, if we denote the first p columns of $\tilde{U}$ by $\tilde{U}_p$, and let $\Lambda_p = \text{diag}(\tilde{\sigma}_1^2, \ldots, \tilde{\sigma}_p^2)$, we have

$$J_2(\tilde{U}_p) = \text{trace}(\tilde{U}_p^T S_m \tilde{U}_p)$$
$$= \text{trace}(\tilde{U}_p^T \tilde{U}_p \Lambda_p)$$
$$= \tilde{\sigma}_1^2 + \cdots + \tilde{\sigma}_p^2$$
$$= \text{trace}(S_m). \tag{1.18}$$

This means that we preserve $\text{trace}(S_m)$ if we take $\tilde{U}_p$ as G. Clearly, the same is true for $\tilde{U}_l$ with $l \geq p$, so PCA to a dimension of at least $\text{rank}(H_m)$ preserves $\text{trace}(S_m)$.

Now we show that LSI also preserves $\text{trace}(S_m)$. Suppose x is an eigenvector of S_m corresponding to the eigenvalue $\lambda \neq 0$. Then

$$S_m x = \sum_{j=1}^{n} (a_j - c)(a_j - c)^T x = \lambda x.$$

This means $x \in \text{span}\{a_j - c | 1 \leq j \leq n\}$, and hence $x \in \text{span}\{a_j | 1 \leq j \leq n\}$. Accordingly,

$$\text{range}(\tilde{U}_p) \subseteq \text{range}(A).$$

From Eq. (1.1), we write

$$A = U_q \Sigma_q V_q^T \quad \text{for} \quad q = \text{rank}(A), \tag{1.19}$$

where U_q and V_q denote the first q columns of U and V, respectively, and $\Sigma_q = \Sigma(1:q, 1:q)$. Then range$(A) = $ range(U_q), which implies that

$$\text{range}(\tilde{U}_p) \subseteq \text{range}(U_q).$$

Hence

$$\tilde{U}_p = U_q W$$

for some matrix $W \in \mathbb{R}^{q \times p}$ with orthonormal columns. This yields

$$
\begin{aligned}
J_2(\tilde{U}_p) &= J_2(U_q W) \\
&= \text{trace}(W^T U_q^T S_m U_q W) \\
&\leq \text{trace}(U_q^T S_m U_q) \\
&= J_2(U_q).
\end{aligned}
$$

Since $J_2(\tilde{U}_p) = $ trace(S_m) from Eq. (1.18), we preserve trace(S_m) if we take U_q as G. The same argument holds for U_l with $l \geq q$, so LSI to any dimension greater than or equal to rank(A) also preserves trace(S_m).

Finally, in the range of reduced dimensions for which PCA and LSI preserve trace(S_m), they preserve trace(S_w) and trace(S_b) as well. This follows from the scatter matrix relationship in Eq. (1.6) and the inequalities

$$
\begin{aligned}
\text{trace}(G^T S_w G) &\leq \text{trace}(S_w) \\
\text{trace}(G^T S_b G) &\leq \text{trace}(S_b),
\end{aligned}
$$

which are satisfied for any G with orthonormal columns, since S_w and S_b are positive semidefinite. In summary, the individual traces of S_m, S_w, and S_b are preserved by using PCA to reduce to a dimension of at least rank(H_m), or by using LSI to reduce to a dimension of at least rank(A).

1.4.2 LSI Plus LDA

In this section, we establish the equivalence of the LDA/GSVD method to a two-stage approach composed of LSI followed by LDA, and denoted by LSI + LDA. Using the notation of Eq. (1.19), the q-dimensional representation of A after the LSI stage is

$$B = U_q^T A,$$

and the second stage applies LDA to B. Letting the superscript B denote matrices after the LSI stage, we have

$$H_b^B = U_q^T H_b \quad \text{and} \quad H_w^B = U_q^T H_w.$$

Hence

$$S_b^B = U_q^T H_b H_b^T U_q \quad \text{and} \quad S_w^B = U_q^T H_w H_w^T U_q.$$

Suppose

$$S_b^B x = \lambda S_w^B x;$$

i.e., x and λ are an eigenvector-eigenvalue pair of the generalized eigenvalue problem that LDA solves in the second stage. Then, for $\lambda = \alpha^2/\beta^2$,

$$\beta^2 U_q^T H_b H_b^T U_q x = \alpha^2 U_q^T H_w H_w^T U_q x.$$

Suppose the matrix (U_q, U_q') is orthogonal. Then $(U_q')^T A = (U_q')^T U_q \Sigma_q V_q^T = 0$, and accordingly, $(U_q')^T H_b = 0$ and $(U_q')^T H_w = 0$, since the columns of both H_b and H_w are linear combinations of the columns of A. Hence

$$\beta^2 \begin{pmatrix} U_q^T \\ (U_q')^T \end{pmatrix} H_b H_b^T U_q x = \begin{pmatrix} \beta^2 U_q^T H_b H_b^T U_q x \\ 0 \end{pmatrix}$$

$$= \begin{pmatrix} \alpha^2 U_q^T H_w H_w^T U_q x \\ 0 \end{pmatrix}$$

$$= \alpha^2 \begin{pmatrix} U_q^T \\ (U_q')^T \end{pmatrix} H_w H_w^T U_q x,$$

which implies

$$\beta^2 H_b H_b^T (U_q x) = \alpha^2 H_w H_w^T (U_q x).$$

That is, $U_q x$ and α/β are a generalized singular vector and value of the generalized singular value problem that LDA solves when applied to A. To show that these $U_q x$ vectors include *all* the LDA solution vectors for A, we show that $\mathrm{rank}(S_m^B) = \mathrm{rank}(S_m)$. From the definition in Eq. (1.11), we have

$$H_m = A - ce^T = A(I - \frac{1}{n}ee^T) = U_q \Sigma_q V_q^T (I - \frac{1}{n}ee^T)$$

and

$$H_m^B = U_q^T H_m,$$

and hence

$$H_m = U_q H_m^B.$$

Since H_m and H_m^B have the same null space, their ranks are the same. This means that the number of non-arbitrary generalized singular value pairs is the same for LDA/GSVD applied to B, which produces $t = \mathrm{rank}(S_m^B)$ pairs, and LDA/GSVD applied to A, which produces $t = \mathrm{rank}(S_m)$ pairs.

We have shown the following.

Theorem 1.4.1 *If G is an optimal LDA transformation for B, the q-dimensional representation of the matrix A via LSI, then $U_q G$ is an optimal LDA transformation for A.*

In other words, LDA applied to A produces

$$Y = (U_q G)^T A = G^T U_q^T A = G^T B,$$

which is the same result as applying LSI to reduce the dimension to q, followed by LDA. Finally, we note that if the dimension after the LSI stage is at least $\mathrm{rank}(A)$, that is $B = U_l^T A$ for $l \geq q$, the equivalency argument remains unchanged.

1.4.3 PCA Plus LDA

As in the previous section for LSI, it can be shown that a two-stage approach in which PCA is followed by LDA is equivalent to LDA applied directly to A. From Eq. (1.17), we write

$$H_m = \tilde{U}_p \tilde{\Sigma}_p \tilde{V}_p^T \quad \text{for} \quad p = \text{rank}(H_m), \qquad (1.20)$$

where $\tilde{U}_p$ and $\tilde{V}_p$ denote the first p columns of $\tilde{U}$ and $\tilde{V}$, respectively, and $\tilde{\Sigma}_p = \tilde{\Sigma}(1:p, 1:p)$. Then the p-dimensional representation of A after the PCA stage is

$$B = \tilde{U}_p^T A,$$

and the second stage applies LDA/GSVD to B. Letting the superscript B denote matrices after the PCA stage, we have

$$S_m^B = \tilde{U}_p^T S_m \tilde{U}_p = \tilde{\Sigma}_p^2, \qquad (1.21)$$

which implies LDA/GSVD applied to B produces $\text{rank}(S_m^B) = p$ non-arbitrary generalized singular value pairs. That is the same number of non-arbitrary pairs as LDA/GSVD applied to A.

We have the following, which is proven in [HP04].

Theorem 1.4.2 *If G is an optimal LDA transformation for B, the p-dimensional representation of the matrix A via PCA, then $\tilde{U}_p G$ is an optimal LDA transformation for A.*

In other words, LDA applied to A produces

$$Y = (\tilde{U}_p G)^T A = G^T \tilde{U}_p^T A = G^T B,$$

which is the same result as applying PCA to reduce the dimension to p, followed by LDA. Note that if the dimension after the PCA stage is at least $\text{rank}(H_m)$, that is $B = \tilde{U}_l^T A$ for $l \geq p$, the equivalency argument remains unchanged.

An additional consequence of Eq. (1.21) is that

$$\text{null}(S_m^B) = \{0\}.$$

Due to the relationship in Eq. (1.6) and the fact that S_w and S_b are positive semidefinite,

$$\text{null}(S_m^B) = \text{null}(S_w^B) \cap \text{null}(S_b^B).$$

Thus the PCA stage eliminates only the joint null space, as illustrated in Figure 1.1(b), which is why we don't lose any discriminatory information before applying LDA.

1.4.4 QRD Plus LDA

To simplify the computation in the first stage, we use the reduced QR decomposition [GV96]

$$A = QR,$$

where $Q \in \mathbb{R}^{m \times n}$ and $R \in \mathbb{R}^{n \times n}$, and let Q play the role that U_q or $\tilde{U}_p$ played before. Then the n-dimensional representation of A after the QR decomposition (QRD) stage is

$$B = Q^T A,$$

and the second stage applies LDA to B. An argument similar to that for LSI [HP04] yields Theorem 1.4.3.

Theorem 1.4.3 *If G is an optimal LDA transformation for B, the n-dimensional representation of the matrix A after QRD, then QG is an optimal LDA transformation for A.*

In other words, LDA applied to A produces

$$Y = (QG)^T A = G^T Q^T A = G^T B,$$

which is the same result as applying QRD to reduce the dimension to n, followed by LDA.

1.5 Factor Analysis Approximations

In this section, we investigate the use of the centroid method as the first step of a two-step process. By using a low-cost SVD approximation, we can avoid truncation and reduce no further than the theoretically optimal intermediate reduced dimension. That is, the centroid approximation may be both inexpensive and accurate enough to outperform an expensive SVD approximation that loses discriminatory information by truncation.

Thurstone [Thu35] gives a complete description of the centroid method, in which he applies the Wedderburn rank reduction process in Eq. (1.4) to the correlation matrix $R = AA^T$. To approximate the SVD, a sign vector (for which each component is 1 or -1) x is chosen so that triple product $x^T R x$ is maximized. This is analogous to finding a general unit vector in which the triple product is maximized. At the kth step, a single factor loading vector is solved for at a time, starting with $x_k = (1 \cdots 1)^T$. The algorithm changes the sign of the element in x_k that increases $x_k^T R_k x_k$ the most, and repeats until any sign change would decrease $x_k^T R_k x_k$.

The rank-one reduction formula is

$$R_{k+1} = R_k - \left(\frac{R_k x_k}{\sqrt{l_k}} \right) \left(\frac{R_k x_k}{\sqrt{l_k}} \right)^T$$

where $l_k = x_k^T R_k x_k$ is the triple product. If $\text{rank}(R) = r$, then a recursion yields

$$R = [R_1 v_1 \cdots R_r v_r] \begin{pmatrix} \frac{1}{l_1} & & \\ & \ddots & \\ & & \frac{1}{l_r} \end{pmatrix} \begin{pmatrix} v_1^T R_1 \\ \vdots \\ v_r^T R_r \end{pmatrix}$$

$$= \left[\frac{R_1 v_1}{\sqrt{l_1}} \cdots \frac{R_r v_r}{\sqrt{l_r}} \right] \begin{pmatrix} \frac{v_1^T R_1}{\sqrt{l_1}} \\ \vdots \\ \frac{v_r^T R_r}{\sqrt{l_r}} \end{pmatrix}.$$

In factor analysis, $\frac{R_k x_k}{\sqrt{l_k}}$ is called the kth factor loading vector.

In [Hor65], the centroid method is described for the data matrix itself. That is, to approximate the SVD of A, sign vectors y and x are chosen so that the bilinear form $y^T A x$ is maximized. At the kth step, the method starts with $x_k = y_k = (1 \cdots 1)^T$. It alternates between changing the sign of the element in y_k that increases $y_k^T A_k x_k$ most, and changing the sign of element in x_k that increases it most. After repeating until any sign change would decrease $y_k^T A_k x_k$, this process yields

$$A = \sum (A_k x_k)(y_k^T A_k x_k)^{-1}(y_k^T A_k),$$

where $(y_k^T A_k x_k)^{-1}$ is split so that $y_k^T A_k$ is normalized.

Chu and Funderlic [CF02] give an algorithm for factoring the correlation matrix AA^T without explicitly forming a cross product. That is, they approximate SVD of AA^T by maximizing $x^T AA^T x$ over sign vectors x. Their algorithm uses pre-factor x_k and post-factor $A_k^T x_k$ as follows:

$$A_{k+1} = A_k - (A_k(A_k^T x_k))(x_k^T A_k(A_k^T x_k))^{-1}(x_k^T A_k).$$

This yields

$$A = \sum (A_k \frac{A_k^T x_k}{\|A_k^T x_k\|})(\frac{x_k^T A_k}{\|A_k^T x_k\|})$$

They also claim that if truncated, the approximation loses statistical meaning unless the rows of A are centered at 0. Finally, they show that the cost of computing l terms of the centroid decomposition involves $\mathcal{O}(lm^2 n)$ complexity for an $m \times n$ data matrix A.

Our goal is to determine how effectively the centroid method approximates the SVD when used as a first stage before applying LDA/GSVD. Toward that end, we have initially implemented the centroid method as applied to the data matrix. To further reduce the computational complexity of the first stage approximation, we will also implement the implicit algorithm of Chu and Funderlic.

1.6 Document Classification Experiments

The first set of experiments were performed on five categories of abstracts from the MEDLINE[2] database. Each category has 500 documents. The dataset was divided

---

[2] http://www.ncbi.nlm.nih.gov/PubMed

Table 1.1. MEDLINE training data set

Class	Category	No. of documents
1	heart attack	250
2	colon cancer	250
3	diabetes	250
4	oral cancer	250
5	tooth decay	250
	dimension	22095×1250

Table 1.2. Classification accuracy (%) on MEDLINE test data

Classification methods	Full 22095×1250	Dimension reduction methods		
		LSI→ 1246 1246×1250	LSI→ 5 5×1250	LDA/GSVD 4×1250
Centroid (L_2)	85.2	85.2	71.6	88.7
Centroid (cosine)	88.3	88.3	78.5	83.9
5NN (L_2)	79.0	79.0	77.8	81.5
15NN (L_2)	83.4	83.4	77.5	88.7
30NN (L_2)	83.8	83.8	77.5	88.7
5NN (cosine)	77.8	77.8	77.8	83.8
15NN (cosine)	82.5	82.5	80.2	83.8
30NN (cosine)	83.8	83.8	79.8	83.8

into 1250 training documents and 1250 test documents (see Table 1.1). After stemming and removal of stop words [Kow97], the training set contains 22,095 distinct terms. Since the dimension (22,095) exceeds the number of training documents (1250), S_w is singular and classical discriminant analysis breaks down. However, LDA/GSVD circumvents this singularity problem.

Table 1.2 reports classification accuracy for the test documents in the full space as well as those in the reduced spaces obtained by LSI and LDA/GSVD methods. Here we use a centroid-based classification method [PJR03], which assigns a document to the cluster to whose centroid it is closest, and K nearest neighbor classification [TK99] for three different values of K. Closeness is determined by both the L_2 norm and cosine similarity measures.

Since the training set has the nearly full rank of 1246, we use LSI to reduce to that. As expected, we observe that the classification accuracies match those from the full space. To illustrate the effectiveness of the GSVD extension, whose optimal reduced dimension is four, LSI reduction to dimension five is included here. With the exception of centroid-based classification using the cosine similarity measure, LDA/GSVD results also compare favorably to those in the original full space, while achieving a significant reduction in time and space complexity. For details, see [KHP05].

To confirm our theoretical results regarding equivalent two-stage methods, we use a MEDLINE dataset of five categories of abstracts with 40 documents in each.

Table 1.3. Traces and classification accuracy (%) on 200 MEDLINE documents

Traces & classification methods	Dimension reduction methods			
	Full 7519×200	LSI 198×200	PCA 197×200	QRD 200×200
Trace(S_w)	73048	73048	73048	73048
Trace(S_b)	6229	6229	6229	6229
Centroid (L_2)	95%	95%	95%	95%
1NN (L_2)	60%	60%	60%	59%
3NN (L_2)	49%	48%	49%	48%

Table 1.4. Traces and classification accuracy (%) on 200 MEDLINE documents

Traces & classification methods	Two-stage methods				
	LSI→ 198 + LDA/GSVD 4×200	PCA→ 197 + LDA/GSVD 4×200	QRD→ 200 + LDA/GSVD 4×200	Centroid→ 198 + LDA/GSVD 4×200
Trace(S_w)	0.05	0.05	0.05	0.05	0.05
Trace(S_b)	3.95	3.95	3.95	3.95	3.95
Centroid (L_2)	99%	99%	99%	99%	99%
1NN (L_2)	99%	99%	99%	99%	99%
3NN (L_2)	98.5%	98.5%	98.5%	99%	98.5%

There are 7519 terms after preprocessing with stemming and removal of stop words [Kow97]. Since 7519 exceeds the number of documents (200), S_w is singular, and classical discriminant analysis breaks down. However, LDA/GSVD and the equivalent two-stage methods circumvent this singularity problem.

Table 1.3 confirms the preservation of the traces of individual scatter matrices upon dimension reduction by the methods we use in the first stage. Specifically, since rank(A) = 198, using LSI to reduce the dimension to 198 preserves the values of trace(S_w) and trace(S_b) from the full space. Likewise, PCA reduction to rank(H_m) = 197 and QRD reduction to n = 200 preserve the individual traces. The effect of these first stages is further illustrated by the lack of significant differences in classification accuracies resulting from each method, as compared to the full space. Closeness is determined by L_2 norm or Euclidean distance.

To confirm the equivalence of the two-stage methods to single-stage LDA/GSVD, we report trace values and classification accuracies for these in Table 1.4. Since S_w is singular, we cannot compute trace($S_w^{-1}S_b$) of the J_1 criterion. However, we observe that trace(S_w) and trace(S_b) are identical for LDA/GSVD and each two-stage method, and they sum to the final reduced dimension of $k - 1 = 4$. Classification results after dimension reduction by each method do not differ significantly, whether obtained by centroid-based or KNN classification.

Finally, the last column in Table 1.4 illustrates how effectively the centroid method approximates the SVD when used as a first stage before LDA/GSVD.

1.7 Conclusion

Our experimental results verify that maximizing the J_1 criterion in Eq. (1.8) effectively optimizes document classification in the reduced-dimensional space, while LDA/GSVD extends its applicability to text data for which S_w is singular. In addition, the LDA/GSVD algorithm avoids the numerical problems inherent in explicitly forming the scatter matrices.

In terms of computational complexity, the most expensive part of Algorithm LDA/GSVD is step 2, where a complete orthogonal decomposition is needed. Assuming $k \leq n$, $t \leq m$, and $t = \mathcal{O}(n)$, the complete orthogonal decomposition of K costs $\mathcal{O}(nmt)$ when $m \leq n$, and $\mathcal{O}(m^2 t)$ when $m > n$. Therefore, a fast algorithm needs to be developed for step 2.

Since $K \in \mathbb{R}^{(k+n) \times m}$, one way to lower the computational cost of LDA/GSVD is to first use another method to reduce the dimension of a document vector from m to n, so that the data matrix becomes a roughly square matrix. For this reason, it is significant that the single-stage LDA/GSVD is equivalent to two-stage methods that use either LSI or PCA as a first stage. Either of these maximizes $J_2(G) = \text{trace}(G^T S_m G)$ over all G with $G^T G = I$, preserving $\text{trace}(S_w)$ and $\text{trace}(S_b)$. The same can be accomplished with the computationally simpler QRD. Thus we provide both theoretical and experimental justification for the increasingly common approach of either LSI + LDA or PCA + LDA, although most studies have reduced the intermediate dimension below that required for equivalence.

Regardless of which approach is taken in the first stage, LDA/GSVD provides both a method for circumventing the singularity that occurs in the second stage and a mathematical framework for understanding the singular case. When applied to the reduced representation in the second stage, the solution vectors correspond one-to-one with those obtained using the single-stage LDA/GSVD. Hence the second stage is a straightforward application of LDA/GSVD to a smaller representation of the original data matrix. Given the relative expense of LDA/GSVD and the two-stage methods, we observe that, in general, QRD is a significantly cheaper first stage for LDA/GSVD than either LSI or PCA. However, if $\text{rank}(A) \ll n$, LSI may be cheaper than the reduced QR decomposition, and will avoid the centering of the data required in PCA. Therefore, the appropriate two-stage method provides a faster algorithm for LDA/GSVD.

We have also proposed a two-stage approach that combines the theoretical advantages of linear discriminant analysis with the computational advantages of factor analysis methods. Here we use the centroid method from factor analysis in the first stage. The motivation stems from its ability to approximate the SVD while simplifying the computational steps. Factor analysis approximations also have the potential to preserve sparsity of the data matrix by restricting the domain of vectors to consider in rank reduction to sign or binary vectors. Our experiments show that the centroid method may provide a sufficiently accurate SVD approximation for the purposes of dimension reduction.

Finally, it bears repeating that dimension reduction is only a preprocessing stage. Since classification and document retrieval will be the dominating parts computationally, the expense of dimension reduction should be weighed against its effectiveness in reducing the cost involved in those processes.

Acknowledgment

This work was supported in part by a New Faculty Research Grant from the Vice President for Research, Utah State University.

References

[BDO95] M.W. Berry, S.T. Dumais, and G.W. O'Brien. Using linear algebra for intelligent information retrieval. *SIAM Review*, 37(4):573–595, 1995.

[BHK97] P.N. Belhumeur, J.P. Hespanha, and D.J. Kriegman. Eigenfaces vs. fisherfaces: recognition using class specific linear projection. *IEEE Transactions on Pattern Analysis and Machine Intelligence*, 19(7):711–720, 1997.

[Bjö96] Å. Björck, *Numerical Methods for Least Squares Problems*. SIAM, 1996.

[CF79] R.E. Cline and R.E. Funderlic. The rank of a difference of matrices and associated generalized inverses. *Linear Algebra Appl.*, 24:185–215, 1979.

[CF02] M.T. Chu and R.E. Funderlic. The centroid decomposition: relationships between discrete variational decompositions and svd. *SIAM J. Matrix Anal. Appl.*, 23:1025–1044, 2002.

[CFG95] M.T. Chu, R.E. Funderlic, and G.H. Golub. A rank-one reduction formula and its applications to matrix factorizations. *SIAM Review*, 37(4):512–530, 1995.

[DDF⁺90] S. Deerwester, S. Dumais, G. Furnas, T. Landauer, and R. Harshman. Indexing by latent semantic analysis. *Journal of the American Society for Information Science*, 41(6):391–407, 1990.

[DHS01] R. Duda, P. Hart, and D. Stork. *Pattern Classification*. John Wiley & Sons, Inc., New York, second edition, 2001.

[Fuk90] K. Fukunaga. *Introduction to Statistical Pattern Recognition*. Academic Press, Boston, second edition, 1990.

[Gut57] L. Guttman. A necessary and sufficient formula for matric factoring. *Psychometrika*, 22(1):79–81, 1957.

[GV96] G. Golub and C. Van Loan. *Matrix Computations*. John Hopkins University Press, Baltimore, MD, third edition, 1996.

[Har67] H.H. Harman. *Modern Factor Analysis*. University of Chicago Press, second edition, 1967.

[HJP03] P. Howland, M. Jeon, and H. Park. Structure preserving dimension reduction for clustered text data based on the generalized singular value decomposition. *SIAM J. Matrix Anal. Appl.*, 25(1):165–179, 2003.

[HMH00] L. Hubert, J. Meulman, and W. Heiser. Two purposes for matrix factorization: a historical appraisal. *SIAM Review*, 42(1):68–82, 2000.

[Hor65] P. Horst. *Factor Analysis of Data Matrices*. Holt, Rinehart and Winston, Inc., 1965.

[HP04] P. Howland and H. Park. Equivalence of several two-stage methods for linear discriminant analysis. In *Proceedings of Fourth SIAM International Conference on Data Mining*, 2004.

[JD88]   A.K. Jain and R.C. Dubes. *Algorithms for Clustering Data*. Prentice Hall, Engle-
           wood Cliffs, NJ, 1988.
[KHP05]   H. Kim, P. Howland, and H. Park. Dimension reduction in text classification with
           support vector machines. *Journal of Machine Learning Research*, 6:37–53, 2005.
[Kow97]   G. Kowalski. *Information Retrieval Systems : Theory and Implementation*. Kluwer
           Academic Publishers, Boston, 1997.
[LH95]   C.L. Lawson and R.J. Hanson. *Solving Least Squares Problems*. SIAM, 1995.
[Loa76]   C.F. Van Loan. Generalizing the singular value decomposition. *SIAM J. Numer.
           Anal.*, 13(1):76–83, 1976.
[Ort87]   J. Ortega. *Matrix Theory: A Second Course*. Plenum Press, New York, 1987.
[PJR03]   H. Park, M. Jeon, and J.B. Rosen. Lower dimensional representation of text data
           based on centroids and least squares. *BIT Numer. Math.*, 42(2):1–22, 2003.
[PS81]   C.C. Paige and M.A. Saunders. Towards a generalized singular value decomposi-
           tion. *SIAM J. Numer. Anal.*, 18(3):398–405, 1981.
[Sal71]   G. Salton. *The SMART Retrieval System*. Prentice-Hall, Englewood Cliffs, NJ,
           1971.
[SM83]   G. Salton and M.J. McGill. *Introduction to Modern Information Retrieval*.
           McGraw-Hill, New York, 1983.
[SW96]   D.L. Swets and J. Weng. Using discriminant eigenfeatures for image retrieval.
           IEEE Transactions on Pattern Analysis and Machine Intelligence, 18(8):831–836,
           1996.
[Thu35]   L.L. Thurstone. *The Vectors of Mind: Multiple Factor Analysis for the Isolation of
           Primary Traits*. University of Chicago Press, Chicago, 1935.
[TK99]   S. Theodoridis and K. Koutroumbas. *Pattern Recognition*. Academic Press, 1999.
[Tor01]   K. Torkkola. Linear discriminant analysis in document classification. In *IEEE
           ICDM Workshop on Text Mining*, San Diego, 2001.
[Wed34]   J.H.M. Wedderburn. *Lectures on Matrices, Colloquium Publications*, volume 17.
           American Mathematical Society, New York, 1934.
[YY03]   J. Yang and J.Y. Yang. Why can LDA be performed in PCA transformed space?
           Pattern Recognition, 36(2):563–566, 2003.

2

Automatic Discovery of Similar Words

Pierre Senellart and Vincent D. Blondel

Overview

The purpose of this chapter is to review some methods used for automatic extraction of similar words from different kinds of sources: large corpora of documents, the World Wide Web, and monolingual dictionaries. The underlying goal of these methods is in general the automatic discovery of synonyms. This goal, however, is most of the time too difficult to achieve since it is often hard to distinguish in an automatic way among synonyms, antonyms, and, more generally, words that are semantically close to each others. Most methods provide words that are "similar" to each other, with some vague notion of semantic similarity. We mainly describe two kinds of methods: techniques that, upon input of a word, automatically compile a list of good synonyms or near-synonyms, and techniques that generate a thesaurus (from some source, they build a complete lexicon of related words). They differ because in the latter case, a complete thesaurus is generated at the same time while there may not be an entry in the thesaurus for each word in the source. Nevertheless, the purposes of both sorts of techniques are very similar and we shall therefore not distinguish much between them.

2.1 Introduction

There are many applications of methods for extracting similar words. For example, in natural language processing and information retrieval, they can be used to broaden and rewrite natural language queries. They can also be used as a support for the compilation of synonym dictionaries, which is a tremendous task. In this chapter we focus on the search of similar words rather than on applications of these techniques.

Many approaches for the automatic construction of thesauri from large corpora have been proposed. Some of them are presented in Section 2.2. The interest of such domain-specific thesauri, as opposed to general-purpose human-written synonym dictionaries, will be stressed. The question of how to combine the result of different techniques will also be broached. We then look at the particular case of the

World Wide Web, whose large size and other specific features do not allow it to be
dealt with in the same way as more classical corpora. In Section 2.3, we propose
an original approach, which is based on a monolingual dictionary and uses an algorithm that generalizes an algorithm initially proposed by Kleinberg for searching the
Web. Two other methods working from a monolingual dictionary are also presented.
Finally, in light of this example technique, we discuss the more fundamental relations that exist between text mining and graph mining techniques for the discovery
of similar words.

2.2 Discovery of Similar Words from a Large Corpus

Much research has been carried out about the search for similar words in textual
corpora, mostly for applications in information retrieval tasks. The basic assumption
of most of these approaches is that words are similar if they are used in the same
contexts. The methods differ in the way the contexts are defined (the document, a
textual window, or more or less elaborate grammatical contexts) and the way the
similarity function is computed.

Depending on the type of corpus, we may obtain different emphasis in the resulting lists of synonyms. The thesaurus built from a corpus is domain-specific to
this corpus and is thus more adapted to a particular application in this domain than
a general human-written dictionary. There are several other advantages to the use of
computer-written thesauri. In particular, they may be rebuilt easily to mirror a change
in the collection of documents (and thus in the corresponding field), and they are not
biased by the lexicon writer (but are of course biased by the corpus in use). Obviously, however, human-written synonym dictionaries are bound to be more liable,
with fewer gross mistakes. In terms of the two classical measures of information retrieval, we expect computer-written thesauri to have a better *recall* (or coverage) and
a lower *precision* (except for words whose meaning is highly biased by the application domain) than general-purpose human-written synonym dictionaries.

We describe below three methods that may be used to discover similar words. We
do not pretend to be exhaustive, but have rather chosen to present some of the main
approaches, selected for the variety of techniques used and specific intents. Variants
and related methods are briefly discussed where appropriate. In Section 2.2.1, we
present a straightforward method, involving a document vector space model and the
cosine similarity measure. This method is used by Chen and Lynch to extract information from a corpus on East-bloc computing [CL92] and we briefly report their
results. We then look at an approach proposed by Crouch [Cro90] for the automatic
construction of a thesaurus. The method is based on a term vector space model and
term discrimination values [SYY75], and is specifically adapted for words that are
not too frequent. In Section 2.2.3, we focus on Grefenstette's SEXTANT system
[Gre94], which uses a partial syntactical analysis. We might need a way to combine the result of various different techniques for building thesauri: this is the object
of Section 2.2.4, which describes the *ensemble* method. Finally, we consider the

particular case of the World Wide Web as a corpus, and discuss the problem of finding synonyms in a very large collection of documents.

2.2.1 A Document Vector Space Model

The first obvious definition of similarity with respect to a context is, given a collection of documents, to say that terms are similar if they tend to occur in the same documents. This can be represented in a multidimensional space, where each document is a dimension and each term is a vector in the document space with boolean entries indicating whether the term appears in the corresponding document. It is common in text mining to use this type of vector space model. In the dual model, terms are coordinates and documents are vectors in term space; we see an application of this dual model in the next section.

Thus, two terms are similar if their corresponding vectors are close to each other. The similarity between the vector $\mathbf{i}$ and the vector $\mathbf{j}$ is computed using a similarity measure, such as *cosine*:

$$\cos(\mathbf{i}, \mathbf{j}) = \frac{\mathbf{i} \cdot \mathbf{j}}{\sqrt{\mathbf{i} \cdot \mathbf{i} \times \mathbf{j} \cdot \mathbf{j}}}$$

where $\mathbf{i} \cdot \mathbf{j}$ is the inner product of $\mathbf{i}$ and $\mathbf{j}$. With this definition we have $|\cos(\mathbf{i}, \mathbf{j})| \leq 1$, defining an angle θ with $\cos \theta = \cos(\mathbf{i}, \mathbf{j})$ as the angle between $\mathbf{i}$ and $\mathbf{j}$. Similar terms tend to occur in the same documents and the angle between them is small (they tend to be *collinear*). Thus, the cosine similarity measure is close to ± 1. On the contrary, terms with little in common do not occur in the same documents, the angle between them is close to $\pi/2$ (they tend to be *orthogonal*), and the cosine similarity measure is close to zero.

Cosine is a commonly used similarity measure. However, one must not forget that the mathematical justification of its use is based on the assumption that the axes are orthogonal, which is seldom the case in practice since documents in the collection are bound to have something in common and not be completely independent.

Chen and Lynch compare in [CL92] the cosine measure with another measure, referred to as the *cluster* measure. The cluster measure is asymmetrical, thus giving asymmetrical similarity relationships between terms. It is defined by:

$$\mathrm{cluster}(\mathbf{i}, \mathbf{j}) = \frac{\mathbf{i} \cdot \mathbf{j}}{\|\mathbf{i}\|_1}$$

where $\|\mathbf{i}\|_1$ is the sum of the magnitudes of $\mathbf{i}$'s coordinates (i.e., the l_1-norm of $\mathbf{i}$).

For both these similarity measures the algorithm is then straightforward: Once a similarity measure has been selected, its value is computed between every pair of terms, and the best similar terms are kept for each term.

The corpus Chen and Lynch worked on was a 200-MB collection of various text documents on computing in the former East-bloc countries. They did not run the algorithms on the raw text. The whole database was manually annotated so that every document was assigned a list of appropriate keywords, countries, organization names, journal names, person names, and folders. Around $60,000$ terms were obtained in this way and the similarity measures were computed on them.

For instance, the best similar keywords (with the cosine measure) for the keyword *technology transfer* were: *export controls*, *trade*, *covert*, *export*, *import*, *micro-electronics*, *software*, *microcomputer*, and *microprocessor*. These are indeed related (in the context of the corpus) and words like *trade*, *import*, and *export* are likely to be some of the best near-synonyms in this context.

The two similarity measures were compared on randomly chosen terms with lists of words given by human experts in the field. Chen and Lynch report that the cluster algorithm presents a better recall (that is, the proportion of relevant terms that are selected) than cosine and human experts. Both similarity measures exhibit similar precisions (that is, the proportion of selected terms that are relevant), which are inferior to that of human experts, as expected. The asymmetry of the cluster measure here seems to be a real advantage.

2.2.2 A Thesaurus of Infrequent Words

Crouch presents in [Cro90] a method for the automatic construction of a thesaurus, consisting of classes of similar words, with only words appearing seldom in the corpus. Her purpose is to use this thesaurus to rewrite queries asked to an information retrieval system. She uses a term vector space model, which is the dual of the space used in previous section: Words are dimensions and documents are vectors. The projection of a vector along an axis is the weight of the corresponding word in the document. Different weighting schemes might be used; one that is effective and widely used is the "term frequency inverse document frequency" (*tf-idf*), that is, the number of times the word appears in the document multiplied by a (monotonous) function of the inverse of the number of documents the word appears in. Terms that appear often in a document and do not appear in many documents have therefore an important weight.

As we saw earlier, we can use a similarity measure such as cosine to characterize the similarity between two vectors (that is, two documents). The algorithm proposed by Crouch, presented in more detail below, is to cluster the set of documents, according to this similarity, and then to select *indifferent discriminators* from the resulting clusters to build thesaurus classes.

Salton, Yang, and Yu introduce in [SYY75] the notion of *term discrimination value*. It is a measure of the effect of the addition of a term (as a dimension) to the vector space on the similarities between documents. A good discriminator is a term that tends to raise the distances between documents; a poor discriminator tends to lower the distances between documents; finally, an indifferent discriminator does not change much the distances between documents. Exact or even approximate computation of all term discrimination values is an expensive task. To avoid this problem, the authors propose to use the term document frequency (i.e., the number of documents the term appears in) instead of the discrimination value, since experiments show they are strongly related. Terms appearing in less than about 1% of the documents are mostly indifferent discriminators; terms appearing in more than 1% and less than 10% of the documents are good discriminators; very frequent terms are poor discriminators. Neither good discriminators (which tend to be specific to subparts of the original corpus) nor poor discriminators (which tend to be stop words or other universally apparent words) are used here.

Crouch suggests using low-frequency terms to form thesaurus classes (these classes should thus be made of indifferent discriminators). The first idea to build the thesaurus would be to cluster together these low-frequency terms with an adequate clustering algorithm. This is not very interesting, however, since, by definition, one has not much information about low-frequency terms. But the documents themselves may be clustered in a meaningful way. The *complete link clustering* algorithm, presented next and which produces small and tight clusters, is adapted to the problem. Each document is first considered as a cluster by itself, and, iteratively, the two closest clusters—the similarity between clusters is defined as the minimum of all similarities (computed by the cosine measure) between pairs of documents in the two clusters—are merged together, until the distance between clusters becomes higher than a user-supplied threshold.

When this clustering step is performed, low-frequency words are extracted from each cluster, thus forming corresponding thesaurus classes. Crouch does not describe these classes but has used them directly for broadening information retrieval queries, and has observed substantial improvements in both recall and precision, on two classical test corpora. It is therefore legitimate to assume that words in the thesaurus classes are related to each other. This method only works on low-frequency words, but the other methods presented here do not generally deal well with such words for which we have little information.

2.2.3 Syntactical Contexts

Perhaps the most successful methods for extracting similar words from text are based on a light syntactical analysis, and the notion of *syntactical context*: For instance, two nouns are similar if they occur as the subject or the direct object of the same verbs. We present here in detail an approach by Grefenstette [Gre94], namely SEXTANT (Semantic EXtraction from Text via Analyzed Networks of Terms); other similar works are discussed next.

Lexical Analysis

Words in the corpus are separated using a simple lexical analysis. A proper name analyzer is also applied. Then, each word is looked up in a human-written lexicon and is assigned a part of speech. If a word has several possible parts of speech, a disambiguator is used to choose the most probable one.

Noun and Verb Phrase Bracketing

Noun and verb phrases are then detected in the sentences of the corpus, using starting, ending, and continuation rules. For instance, a determiner can start a noun phrase, a noun can follow a determiner in a noun phrase, an adjective cannot neither start, end, or follow any kind of word in a verb phrase, and so on.

Parsing

Several syntactic relations (or contexts) are then extracted from the bracketed sentences, requiring five successive passes over the text. Table 2.1, taken from [Gre94], shows the list of extracted relations.

Table 2.1. Syntactical relations extracted by SEXTANT

ADJ	an adjective modifies a noun	(e.g., *civil unrest*)
NN	a noun modifies a noun	(e.g., *animal rights*)
NNPREP	a noun that is the object of a proposition modifies a preceding noun	(e.g., *measurements along the crest*)
SUBJ	a noun is the subject of a verb	(e.g., *the table shook*)
DOBJ	a noun is the direct object of a verb	(e.g., *he ate an apple*)
IOBJ	a noun in a prepositional phrase modifying a verb	(e.g., *the book was placed on the table*)

The relations generated are thus not perfect (on a sample of 60 sentences Grefenstette found a correctness ratio of 75%) and could be better if a more elaborate parser was used, but it would be more expensive too. Five passes over the text are enough to extract these relations, and since the corpus used may be very large, backtracking, recursion or other time-consuming techniques used by elaborate parsers would be inappropriate.

Similarity

Grefenstette focuses on the similarity between nouns; other parts of speech are not dealt with. After the parsing step, a noun has a number of attributes: all the words that modify it, along with the kind of syntactical relation (**ADJ** for an adjective, **NN** or **NNPREP** for a noun and **SUBJ**, **DOBJ**, or **IOBJ** for a verb). For instance, the noun *cause*, which appears 83 times in a corpus of medical abstracts, has 67 unique attributes in this corpus. These attributes constitute the context of the noun, on which similarity computations are made. Each attribute is assigned a weight by:

$$\text{weight}(att) = 1 + \sum_{\text{noun } i} \frac{p_{att,i} \log(p_{att,i})}{\log(\text{total number of relations})}$$

where

$$p_{att,i} = \frac{\text{number of times } att \text{ appears with } i}{\text{total number of attributes of } i}$$

The similarity measure used by Grefenstette is a weighted *Jaccard similarity measure* defined as follows:

$$\mathrm{jac}(i,j) = \frac{\sum_{att \text{ attribute of both } i \text{ and } j} \text{weight}(att)}{\sum_{att \text{ attribute of either } i \text{ or } j} \text{weight}(att)}$$

Results

Table 2.2. SEXTANT similar words for *case*, from different corpora

1. CRAN (Aeronautics abstract)
   case: *characteristic, analysis, field, distribution, flaw, number, layer, problem*
2. JFK (Articles on JFK assassination conspiracy theories)
   case: *film, evidence, investigation, photograph, picture, conspiracy, murder*
3. MED (Medical abstracts)
   case: *change, study, patient, result, treatment, child, defect, type, disease, lesion*

Grefenstette used SEXTANT on various corpora and many examples of the results returned are available in [Gre94]. Table 2.2 shows the most similar words of *case* in three completely different corpora. It is interesting to note that the corpus has a great impact on the meaning of the word according to which similar words are selected. This is a good illustration of the interest of working on a domain-specific corpus.

Table 2.3. SEXTANT similar words for words with most contexts in Grolier's Encyclopedia animal articles

species	*bird, fish, family, group, form, animal, insect, range, snake*
fish	*animal, species, bird, form, snake, insect, group, water*
bird	*species, fish, animal, snake, insect, form, mammal, duck*
water	*sea, area, region, coast, forest, ocean, part, fish, form, lake*
egg	*nest, female, male, larva, insect, day, form, adult*

Table 2.3 shows other examples, in a corpus on animals. Most words are closely related to the initial word and some of them are indeed very good (*sea, ocean, lake* for *water*; *family, group* for *species...*) There remain completely unrelated words though, such as *day* for *egg*.

Other Techniques Based on a Light Syntactical Analysis

A number of works deal with the extraction of similar words from corpora with the help of a light syntactical analysis. They rely on grammatical contexts, which can be seen as 3-tuples (w, r, w'), where w and w' are two words and r characterizes the relation between w and w'. In particular, [Lin98] and [CM02] propose systems quite similar to SEXTANT, and apply them to much larger corpora. Another interesting feature of these works is that the authors try to compare numerous similarity measures; [CM02] especially presents an extensive comparison of the results obtained with different similarity and weight measures.

Another interesting approach is presented in [PTL93]. The *relative entropy* between distributions of grammatical contexts for each word is used as a similarity measure between these two words, and this similarity measure is used in turn for a hierarchical clustering of the set of words. This clustering provides a rich thesaurus of similar words. Only the **DOBJ** relation is considered in [PTL93], but others can be used in the same manner.

2.2.4 Combining the Output of Multiple Techniques

The techniques presented above may use different similarity measures, different parsers, or may have different inherent biases. In some contexts, using a combination of various techniques may be useful to increase the overall quality of lists of similar words. A general solution to this problem in the general context of machine learning is the use of *ensemble* methods [Die00]; these methods may be fairly elaborate, but a simple one (*Bayesian voting*) amounts to performing some renormalization of the similarity scores and averaging them together. Curran uses such a technique in [Cur02] to aggregate the results of different techniques based on a light parsing; each of these uses the same similarity measure, making the renormalization step useless. Another use of the combination of different techniques is to be able to benefit from different kinds of sources: Wu and Zhou [WZ03] extend Curran's approach to derive a thesaurus of similar words from very different sources: a monolingual dictionary (using a method similar to the distance method of Section 2.3.3), a monolingual corpus (using grammatical contexts), and the combination of a bilingual dictionary and a bilingual corpus with an original algorithm.

2.2.5 How to Deal with the Web

The World Wide Web is a very particular corpus: Its size simply cannot be compared with the largest corpora traditionally used for synonym extraction, its access times are high, and it is also richer and more lively than any other corpus. Moreover, a large part of it is conveniently indexed by search engines. One could imagine that its hyperlinked structure could be of some use too (see the discussion in Section 2.3.7). And of course it is not a domain-specific source, though domain-specific parts of the Web could be extracted by restricting ourselves to pages matching appropriate keyword queries. Is it possible to use the Web for the discovery of similar words? Obviously, because of the size of the Web, none of the above techniques can apply.

Turney partially deals with the issue in [Tur01]. He does not try to obtain a list of synonyms of a word i but, given a word i, he proposes a way to assign a synonymy score to any word j. His method was validated against synonym recognition questions extracted from two English tests: the Test Of English as a Foreign Language (TOEFL) and the English as a Second Language test (ESL). Four different synonymy scores are compared, and each of these use the advanced search capabilities of the Altavista search engine (http://www.altavista.com/).

$$\text{score}_1(j) = \frac{\text{hits}(i \text{ AND } j)}{\text{hits}(j)}$$

$$\text{score}_2(j) = \frac{\text{hits}(i \text{ NEAR } j)}{\text{hits}(j)}$$

$$\text{score}_3(j) = \frac{\text{hits}((i \text{ NEAR } j) \text{ AND NOT } ((i \text{ OR } j) \text{ NEAR } not))}{\text{hits}(j \text{ AND NOT}(j \text{ NEAR } not))}$$

$$\text{score}_4(j) = \frac{\text{hits}((i \text{ NEAR } j) \text{ AND } context \text{ AND NOT } ((i \text{ OR } j) \text{ NEAR } not))}{\text{hits}(j \text{ AND } context \text{ AND NOT}(j \text{ NEAR } not))}$$

In these expressions, $\text{hits}(\cdot)$ represents the number of pages returned by Altavista for the corresponding query, AND, OR, and NOT are the classical boolean operators, NEAR imposes that the two words are not separated by more than ten words, and *context* is a context word (a context was given along with the question in ESL, the context word may be automatically derived from it). The difference between score_2 and score_3 was introduced in order not to assign a good score to antonyms.

The four scores are presented in increasing order of quality of the corresponding results: score_3 gives the right synonym for 73.75% of the questions from TOEFL (score_4 was not applicable since no context was given) and score_4 gives the right synonym in 74% of the questions from ESL. These results are arguably good, since, as reported by Turney, the average score of TOEFL by a large sample of students is 64.5%.

This algorithm cannot be used to obtain a global synonym dictionary, as it is too expensive to run for each candidate word in a dictionary because of network access times, but it may be used, for instance, to refine a list of synonyms given by another method.

2.3 Discovery of Similar Words in a Dictionary

2.3.1 Introduction

We propose now a method for automatic synonym extraction in a monolingual dictionary [Sen01]. Our method uses a graph constructed from the dictionary and is based on the assumption that synonyms have many words in common in their definitions and are used in the definition of many common words. Our method is based on an algorithm that generalizes the *HITS* algorithm initially proposed by Kleinberg for searching the Web [Kle99].

Starting from a dictionary, we first construct the associated *dictionary graph G*; each word of the dictionary is a vertex of the graph and there is an edge from u to v if v appears in the definition of u. Then, associated to a given query word w, we construct a *neighborhood graph G_w* that is the subgraph of G whose vertices are those pointed to by w or pointing to w. Finally, we look in the graph G_w for vertices that are similar to the vertex 2 in the structure graph

$$1 \longrightarrow 2 \longrightarrow 3$$

and choose these as synonyms. For this last step we use a similarity measure between vertices in graphs that was introduced in [BGH+04].

The problem of searching synonyms is similar to that of searching similar pages on the Web, a problem that is dealt with in [Kle99] and [DH99]. In these references, similar pages are found by searching authoritative pages in a subgraph focused on the original page. Authoritative pages are pages that are similar to the vertex "authority" in the structure graph

$$\text{hub} \longrightarrow \text{authority}.$$

We ran the same method on the dictionary graph and obtained lists of good hubs and good authorities of the neighborhood graph. There were duplicates in these lists but not all good synonyms were duplicated. Neither authorities nor hubs appear to be the right concept for discovering synonyms.

In the next section, we describe our method in some detail. In Section 2.3.3, we briefly survey two other methods that are used for comparison. We then describe in Section 2.3.4 how we have constructed a dictionary graph from 1913 Webster's dictionary. We compare next the three methods on a sample of words chosen for their variety. Finally, we generalize the approach presented here by discussing the relations existing between the fields of text mining and graph mining, in the context of synonym discovery.

2.3.2 A Generalization of Kleinberg's Method

In [Kle99], Jon Kleinberg proposes the HITS method for identifying Web pages that are good *hubs* or good *authorities* for a given query. For example, for the query "automobile makers," the home pages of Ford, Toyota and other car makers are good authorities, whereas Web pages that list these home pages are good hubs. To identify hubs and authorities, Kleinberg's method exploits the natural graph structure of the Web in which each Web page is a vertex and there is an edge from vertex a to vertex b if page a points to page b. Associated to any given query word w, the method first constructs a "focused subgraph" G_w analogous to our neighborhood graph and then computes hub and authority scores for all vertices of G_w. These scores are obtained as the result of a converging iterative process. Initial hub and authority weights are all set to one, $x^1 = 1$ and $x^2 = 1$. These initial weights are then updated simultaneously according to a *mutually reinforcing rule*: The hub score of the vertex i, x_i^1, is set equal to the sum of the authority scores of all vertices pointed by i and, similarly, the authority scores of the vertex j, x_j^2, is set equal to the sum of the hub scores of all vertices pointing to j. Let M_w be the adjacency matrix associated to G_w. The updating equations can be written as

$$\begin{pmatrix} x^1 \\ x^2 \end{pmatrix}_{t+1} = \begin{pmatrix} 0 & M_w \\ M_w^T & 0 \end{pmatrix} \begin{pmatrix} x^1 \\ x^2 \end{pmatrix}_t \qquad t = 0, 1, \dots$$

It can be shown that under weak conditions the normalized vector x^1 (respectively, x^2) converges to the normalized principal eigenvector of $M_w M_w^T$ (respectively, $M_w^T M_w$).

The authority score of a vertex v in a graph G can be seen as a similarity measure between v in G and vertex 2 in the graph

$$1 \longrightarrow 2.$$

Similarly, the hub score of v can be seen as a measure of similarity between v in G and vertex 1 in the same structure graph. As presented in [BGH+04], this measure of similarity can be generalized to graphs that are different from the authority-hub structure graph. We describe below an extension of the method to a structure graph with three vertices and illustrate an application of this extension to synonym extraction.

Let G be a dictionary graph. The neighborhood graph of a word w is constructed with the words that appear in the definition of w and those that use w in their definition. Because of this, the word w in G_w is similar to the vertex 2 in the structure graph (denoted P_3)

$$1 \longrightarrow 2 \longrightarrow 3.$$

For instance, Figure 2.1 shows a part of the neighborhood graph of *likely*. The words *probable* and *likely* in the neighborhood graph are similar to the vertex 2 in P_3. The words *truthy* and *belief* are similar to, respectively, vertices 1 and 3. We say that a vertex is similar to the vertex 2 of the preceding graph if it points to vertices that are similar to the vertex 3 and if it is pointed to by vertices that are similar to the vertex 1. This mutually reinforcing definition is analogous to Kleinberg's definitions of hubs and authorities.

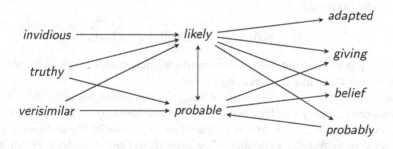

Fig. 2.1. Subgraph of the neighborhood graph of *likely*.

The similarity between vertices in graphs can be computed as follows. To every vertex i of G_w we associate three scores (as many scores as there are vertices in the structure graph) x_i^1, x_i^2, and x_i^3 and initially set them equal to one. We then iteratively update the scores according to the following mutually reinforcing rule: The score x_i^1 is set equal to the sum of the scores x_j^2 of all vertices j pointed by i; the score x_i^2

is set equal to the sum of the scores x_j^3 of vertices pointed by i and the scores x_j^1 of vertices pointing to i; finally, the score x_i^3 is set equal to the sum of the scores x_j^2 of vertices pointing to i. At each step, the scores are updated simultaneously and are subsequently normalized:

$$x^k \leftarrow \frac{x^k}{\|x^k\|} \qquad (k = 1, 2, 3).$$

It can be shown that when this process converges, the normalized vector score x^2 converges to the normalized principal eigenvector of the matrix $M_w M_w^T + M_w^T M_w$. Thus, our list of synonyms can be obtained by ranking in decreasing order the entries of the principal eigenvector of $M_w M_w^T + M_w^T M_w$.

2.3.3 Other Methods

In this section, we briefly describe two synonym extraction methods that will be compared to ours on a selection of four words.

The Distance Method

One possible way of defining a synonym distance is to declare that two words are close to being synonyms if they appear in the definition of many common words and have many common words in their definition. A way of formalizing this is to define a distance between two words by counting the number of words that appear in one of the definitions but not in both, and add to this the number of words that use one of the words but not both in their definition. Let A be the adjacency matrix of the dictionary graph, and i and j be the vertices associated to two words. The distance between i and j can be expressed as

$$d(i, j) = \|(A_{i,\cdot} - A_{j,\cdot})\|_1 + \|(A_{\cdot,i} - A_{\cdot,j})^T\|_1$$

where $\| \cdot \|_1$ is the l_1-norm. For a given word i we may compute $d(i, j)$ for all j and sort the words according to increasing distance.

Unlike the other methods presented here, we can apply this algorithm directly to the entire dictionary graph rather than on the neighborhood graph. However, this gives very bad results: The first two synonyms of *sugar* in the dictionary graph constructed from Webster's Dictionary are *pigwidgeon* and *ivoride*. We shall see in Section 2.3.5 that much better results are achieved if we use the neighborhood graph.

ArcRank

ArcRank is a method introduced by Jannink and Wiederhold for building a thesaurus [JW99]; their intent was not to find synonyms but related words. The method is based on the PageRank algorithm, used by the Web search engine Google and described in [BP98]. PageRank assigns a ranking to each vertex of the dictionary graph in the following way. All vertices start with identical initial ranking and then iteratively distribute it to the vertices they point to, while receiving the sum of the ranks from vertices they are pointed to by. Under conditions that are often satisfied in practice, the normalized ranking converges to a stationary distribution corresponding to the principal eigenvector of the adjacency matrix of the graph. This algorithm is actually slightly modified so that sources (nodes with no incoming edges, that is words not used in any definition) and sinks (nodes with no outgoing edges, that is words not defined) are not assigned extreme rankings.

ArcRank assigns a ranking to each edge according to the ranking of its vertices. If $|a_s|$ is the number of outgoing edges from vertex s and p_t is the page rank of vertex t, then the edge relevance of (s, t) is defined by

$$r_{s,t} = \frac{p_s/|a_s|}{p_t}$$

Edge relevances are then converted into rankings. Those rankings are computed only once. When looking for words related to some word w, one selects the edges starting from or arriving to w that have the best rankings and extract the corresponding incident vertices.

2.3.4 Dictionary Graph

Before proceeding to the description of our experiments, we describe how we constructed the dictionary graph. We used the Online Plain Text English Dictionary [OPT], which is based on the "Project Gutenberg Etext of Webster's Unabridged Dictionary," which is in turn based on the 1913 U.S. Webster's Unabridged Dictionary. The dictionary consists of 27 HTML files (one for each letter of the alphabet, and one for several additions). These files are available from the Website `http://www.gutenberg.net/`. To obtain the dictionary graph, several choices had to be made.

- Some words defined in Webster's dictionary are multi-words (e.g., *All Saints*, *Surinam toad*). We did not include these words in the graph since there is no simple way to decide, when the words are found side-by-side, whether or not they should be interpreted as single words or as a multi-word (for instance, *at one* is defined but the two words *at* and *one* appear several times side-by-side in the dictionary in their usual meanings).
- Some head words of definitions were prefixes or suffixes (e.g., *un-*, *-ous*), these were excluded from the graph.
- Many words have several meanings and are head words of multiple definitions. For, once more, it is not possible to determine which meaning of a word is employed in a definition, we gathered the definitions of a word into a single one.
- The recognition of inflected forms of a word in a definition is also a problem. We dealt with the cases of regular and semiregular plurals (e.g., *daisies*, *albatrosses*) and regular verbs, assuming that irregular forms of nouns or verbs (e.g., *oxen*, *sought*) had entries in the dictionary. Note that a classical stemming here would

not be of use, since we do not want to merge the dictionary entries of lexically close words, such as *connect* and *connection*).

The resulting graph has 112,169 vertices and 1,398,424 edges, and can be downloaded at `http://pierre.senellart.com/stage_maitrise/graphe/`. We analyzed several features of the graph: connectivity and strong connectivity, number of connected components, distribution of connected components, degree distributions, graph diameter, etc. Our findings are reported in [Sen01].

We also decided to exclude stop words in the construction of neighborhood graphs, that is words that appear in more than L definitions (best results were obtained for $L \approx 1,000$).

2.3.5 Results

To be able to compare the different methods presented above (Distance, ArcRank, and our method based on graph similarity) and to evaluate their relevance, we examine the first ten results given by each of them for four words, chosen for their variety.

disappear	a word with various synonyms such as *vanish*.
parallelogram	a very specific word with no true synonyms but with some similar words: *quadrilateral, square, rectangle, rhomb...*
sugar	a common word with different meanings (in chemistry, cooking, dietetics...). One can expect *glucose* as a candidate.
science	a common and vague word. It is hard to say what to expect as synonym. Perhaps *knowledge* is the best option.

Words of the English language belong to different parts of speech: nouns, verbs, adjectives, adverbs, prepositions, etc. It is natural, when looking for a synonym of a word, to get only words of the same kind. Websters's Dictionary provides for each word its part of speech. But this presentation has not been standardized and we counted no less than 305 different categories. We have chosen to select five types: nouns, adjectives, adverbs, verbs, others (including articles, conjunctions, and interjections), and have transformed the 305 categories into combinations of these types. A word may of course belong to different types. Thus, when looking for synonyms, we have excluded from the list all words that do not have a common part of speech with our word. This technique may be applied with all synonym extraction methods but since we did not implement ArcRank, we did not use it for ArcRank. In fact, the gain is not huge, because many words in English have several grammatical natures. For instance, *adagio* or *tete-a-tete* are at the same time nouns, adjectives, and adverbs.

We have also included lists of synonyms coming from WordNet [Wor], which is human-written. The order of appearance of the words for this last source is arbitrary, whereas it is well defined for the distance method and for our method. The results given by the Web interface implementing ArcRank are two rankings, one for words pointed by and one for words pointed to. We have interleaved them into one ranking.

We have not kept the query word in the list of synonyms, since this has not much sense except for our method, where it is interesting to note that in every example we have experimented with, the original word appeared as the first word of the list (a point that tends to give credit to the method).

To have an objective evaluation of the different methods, we asked a sample of 21 persons to give a mark (from 0 to 10, 10 being the best one) to the lists of synonyms, according to their relevance to synonymy. The lists were of course presented in random order for each word. Tables 2.4, 2.5, 2.6, and 2.7 give the results.

Table 2.4. Proposed synonyms for *disappear*

	Distance	Our method	ArcRank	WordNet
1	vanish	vanish	epidemic	vanish
2	wear	pass	disappearing	go away
3	die	die	port	end
4	sail	wear	dissipate	finish
5	faint	faint	cease	terminate
6	light	fade	eat	cease
7	port	sail	gradually	
8	absorb	light	instrumental	
9	appear	dissipate	darkness	
10	cease	cease	efface	
Mark	3.6	6.3	1.2	7.5
Std dev.	1.8	1.7	1.2	1.4

Concerning *disappear*, the distance method (restricted to the neighborhood graph) and our method do pretty well. *vanish, cease, fade, die, pass, dissipate, faint* are very relevant (one must not forget that verbs necessarily appear without their postposition). *dissipate* or *faint* are relevant too. However, some words like *light* or *port* are completely irrelevant, but they appear only in 6th, 7th, or 8th position. If we compare these two methods, we observe that our method is better: An important synonym like *pass* takes a good ranking, whereas *port* or *appear* go out of the top ten words. It is hard to explain this phenomenon, but we can say that the mutually reinforcing aspect of our method is apparently a positive point. On the contrary, Arc-Rank gives rather poor results with words such as *eat, instrumental*, or *epidemic* that are out of the point.

Because the neighborhood graph of *parallelogram* is rather small (30 vertices), the first two algorithms give similar results, which are not absurd: *square, rhomb, quadrilateral, rectangle,* and *figure* are rather interesting. Other words are less relevant but still are in the semantic domain of *parallelogram*. ArcRank, which also works on the same subgraph, does not give as interesting words, although *gnomon* makes its appearance, since *consequently* and *popular* are irrelevant. It is interesting to note that WordNet here is less rich because it focuses on a particular aspect (*quadrilateral*).

Table 2.5. Proposed synonyms for *parallelogram*

	Distance	Our method	ArcRank	WordNet
1	*square*	*square*	*quadrilateral*	*quadrilateral*
2	*parallel*	*rhomb*	*gnomon*	*quadrangle*
3	*rhomb*	*parallel*	*right-lined*	*tetragon*
4	*prism*	*figure*	*rectangle*	
5	*figure*	*prism*	*consequently*	
6	*equal*	*equal*	*parallelepiped*	
7	*quadrilateral*	*opposite*	*parallel*	
8	*opposite*	*angles*	*cylinder*	
9	*altitude*	*quadrilateral*	*popular*	
10	*parallelepiped*	*rectangle*	*prism*	
Mark	4.6	4.8	3.3	6.3
Std dev.	2.7	2.5	2.2	2.5

Table 2.6. Proposed synonyms for *sugar*

	Distance	Our method	ArcRank	WordNet
1	*juice*	*cane*	*granulation*	*sweetening*
2	*starch*	*starch*	*shrub*	*sweetener*
2	*cane*	*sucrose*	*sucrose*	*carbohydrate*
4	*milk*	*milk*	*preserve*	*saccharide*
5	*molasses*	*sweet*	*honeyed*	*organic compound*
6	*sucrose*	*dextrose*	*property*	*saccarify*
7	*wax*	*molasses*	*sorghum*	*sweeten*
8	*root*	*juice*	*grocer*	*dulcify*
9	*crystalline*	*glucose*	*acetate*	*edulcorate*
10	*confection*	*lactose*	*saccharine*	*dulcorate*
Mark	3.9	6.3	4.3	6.2
Std dev.	2.0	2.4	2.3	2.9

Once more, the results given by ArcRank for *sugar* are mainly irrelevant (*property*, *grocer*...) Our method is again better than the distance method: *starch*, *sucrose*, *sweet*, *dextrose*, *glucose*, and *lactose* are highly relevant words, even if the first given near-synonym (*cane*) is not as good. Its given mark is even better than for WordNet.

The results for *science* are perhaps the most difficult to analyze. The distance method and ours are comparable. ArcRank gives perhaps better results than for other words but is still poorer than the two other methods.

As a conclusion, the first two algorithms give interesting and relevant words, whereas it is clear that ArcRank is not adapted to the search for synonyms. The variation of Kleinberg's algorithm and its mutually reinforcing relationship demonstrates its superiority on the basic distance method, even if the difference is not obvious for all words. The quality of the results obtained with these different methods is still quite different from that of human-written dictionaries such as WordNet. Still, these automatic techniques show their interest, since they present more complete aspects

Table 2.7. Proposed synonyms for *science*

	Distance	Our method	ArcRank	WordNet
1	*art*	*art*	*formulate*	*knowledge domain*
2	*branch*	*branch*	*arithmetic*	*knowledge base*
3	*nature*	*law*	*systematize*	*discipline*
4	*law*	*study*	*scientific*	*subject*
5	*knowledge*	*practice*	*knowledge*	*subject area*
6	*principle*	*natural*	*geometry*	*subject field*
7	*life*	*knowledge*	*philosophical*	*field*
8	*natural*	*learning*	*learning*	*field of study*
9	*electricity*	*theory*	*expertness*	*ability*
10	*biology*	*principle*	*mathematics*	*power*
Mark	3.6	4.4	3.2	7.1
Std dev.	2.0	2.5	2.9	2.6

of a word than human-written dictionaries. They can profitably be used to broaden a topic (see the example of *parallelogram*) and to help with the compilation of synonym dictionaries.

2.3.6 Perspectives

A first immediate improvement of our method would be to work on a larger subgraph than the neighborhood subgraph. The neighborhood graph we have introduced may be rather small, and therefore may not include important near-synonyms. A good example is *ox*, of which *cow* seems to be a good synonym. Unfortunately, *ox* does not appear in the definition of *cow*, neither does the latter appear in the definition of the former. Thus, the methods described above cannot find this word. Larger neighborhood graphs could be obtained either as Kleinberg does in [Kle99] for searching similar pages on the Web, or as Dean and Henzinger do in [DH99] for the same purpose. However, such subgraphs are not any longer focused on the original word. That implies that our variation of Kleinberg's algorithm "forgets" the original word and may produce irrelevant results. When we use the vicinity graph of Dean and Henzinger, we obtain a few interesting results with specific words: For example, *trapezoid* appears as a near-synonym of *parallelogram* or *cow* as a near-synonym of *ox*. Yet there are also many degradations of performance for more general words. Perhaps a choice of neighborhood graph that depends on the word itself would be appropriate. For instance, the extended vicinity graph may be used either for words whose neighborhood graph has less than a fixed number of vertices, or for words whose incoming degree is small, or for words who do not belong to the largest connected component of the dictionary graph.

One may wonder whether the results obtained are specific to Webster's dictionary or whether the same methods could work on other dictionaries (using domain-specific dictionaries could for instance generate domain-specific thesauri, whose interest was mentioned in Section 2.2), in English or in other languages. Although the latter is most likely since our techniques were not designed for the particular graph we worked on, there are undoubtedly differences with other languages. For example, in French, postpositions do not exist and thus verbs do not have as many different meanings as in English. Besides, it is much rarer in French to have the same word for the noun and for the verb than in English. Furthermore, the way words are defined vary from language to language. Despite these differences, preliminary studies on a monolingual French dictionary seem to show equally good results.

2.3.7 Text Mining and Graph Mining

All three methods described for synonym extraction from a dictionary use classical techniques from text mining: stemming (in our case, in the form of a simple lemmatization), stop-word removal, a vector space model for representing dictionary entries... But a specificity of monolingual dictionaries makes this vector space very peculiar: Both the dimensions of the vector space and the vectors stand for the same kind of objects—words. In other words, rows and columns of the corresponding matrix are indexed by the same set. This peculiarity makes it possible to see the dictionary, and this vector space model, as a (possibly weighted) directed graph. This allows us to see the whole synonym extraction problem as a problem of information retrieval on graphs, for which a number of different approaches have been elaborated, especially in the case of the World Wide Web [BP98, DH99, Kle99]. Thus, classical techniques from both text mining (distance between vectors, cosine similarity, tf-idf weighting...) and graph mining (cocitation count, PageRank, HITS, graph similarity measures...) can be used in this context. A study [OS07] on the Wikipedia on-line encyclopedia [Wik], which is similar to a monolingual dictionary, compares some methods from both worlds, along with an original approach for defining similarity in graphs based on *Green measures* of Markov chains.

A further step would be to consider *any* text mining problem as a graph mining problem, by considering any finite set of vectors (in a finite-dimensional vector space) as a directed, weighted, bipartite graph, the two partitions representing respectively the vectors and the dimensions. Benefits of this view are somewhat lower, because of the very particular nature of a bipartite graph, but some notions from graph theory (for instance, matchings, vertex covers, or bipartite random walks), may still be of interest.

2.4 Conclusion

A number of different methods exist for the automatic discovery of similar words. Most of these methods are based on various text corpora, and three of these are described in this chapter. Each of them may be more or less adapted to a specific problem (for instance, Crouch's techniques are more adapted to infrequent words than SEXTANT). We have also described the use of a more structured source—a monolingual dictionary—for the discovery of similar words. None of these methods is perfect and in fact none of them favorably competes with human-written dictionaries in terms of liability. Computer-written thesauri, however, have other advantages such as their ease to build and maintain. We also discussed how different methods, with their own pros and cons, might be integrated.

Another problem of the methods presented is the vagueness of the notion of "similar word" that they use. Depending on the context, this notion may or may not include the notion of synonyms, near-synonyms, antonyms, hyponyms, etc. The distinction between these very different notions by automatic means is a challenging problem that should be addressed to make it possible to build thesauri in a completely automatic way.

Acknowledgment

We would like to thank Yann Ollivier for his feedback on this work.

References

[BGH+04] V.D. Blondel, A. Gajardo, M. Heymans, P. Senellart, and P. Van Dooren. A measure of similarity between graph vertices: applications to synonym extraction and
Web searching. *SIAM Review*, 46(4):647–666, 2004.

[BP98] S. Brin and L. Page. The anatomy of a large-scale hypertextual Web search engine.
Computer Networks and ISDN Systems, 30(1-7):107–117, 1998.

[CL92] H. Chen and K.J. Lynch. Automatic construction of networks of concepts characterizing document databases. *IEEE Transactions on Systems, Man and Cybernet-
ics*, 22(5):885–902, 1992.

[CM02] J.R. Curran and M. Moens. Improvements in automatic thesaurus extraction. In
Proc. ACL SIGLEX, Philadelphia, July 2002.

[Cro90] C.J. Crouch. An approach to the automatic construction of global thesauri. *Information Processing and Management*, 26(5):629–640, 1990.

[Cur02] J.R. Curran. Ensemble methods for automatic thesaurus extraction. In *Proc. Conference on Empirical Methods in Natural Language Processing*, Philadelphia, July
2002.

[DH99] J. Dean and M.R. Henzinger. Finding related pages in the world wide web. In
Proc. WWW, Toronto, Canada, May 1999.

[Die00] T.G. Dietterich. Ensemble methods in machine learning. In *Proc. MCS*, Cagliari,
Italy, June 2000.

[Gre94] G. Grefenstette. *Explorations in Automatic Thesaurus Discovery*. Kluwer Academic Press, Boston, MA, 1994.

[JW99] J. Jannink and G Wiederhold. Thesaurus entry extraction from an on-line dictionary. In *Proc. FUSION*, Sunnyvale, CA, July 1999.

[Kle99] J.M. Kleinberg. Authoritative sources in a hyperlinked environment. *Journal of
the ACM*, 46(5):604–632, 1999.

[Lin98] D. Lin. Automatic retrieval and clustering of similar words. In *Proc. COLING*,
Montreal, Canada, August 1998.

[OPT] The online plain text English dictionary. http://msowww.anu.edu.au/
~ralph/OPTED/.

[OS07] Y. Ollivier and P. Senellart. Finding related pages using Green measures: An illustration with Wikipedia. In *Proc. AAAI*, Vancouver, Canada, July 2007.

[PTL93]   F. Pereira, N. Tishby, and L. Lee. Distributional clustering of english words. In *Proc. ACL*, Columbus, OH, June 1993.

[Sen01]   P. Senellart. Extraction of information in large graphs. Automatic search for synonyms. Technical Report 90, Université catholique de Louvain, Louvain-la-neuve, Belgium, 2001.

[SYY75]   G. Salton, C.S. Yang, and C.T. Yu. A theory of term importance in automatic text analysis. *Journal of the American Society for Information Science*, 26(1):33–44, 1975.

[Tur01]   P.D. Turney. Mining the Web for Synonyms: PMI-IR versus LSA on TOEFL. In *Proc. ECML*, Freiburg, Germany, September 2001.

[Wik]   Wikipedia. The free encyclopedia. http://en.wikipedia.org/.

[Wor]   WordNet 1.6. http://wordnet.princeton.edu/.

[WZ03]   H. Wu and M. Zhou. Optimizing synonym extraction using monolingual and bilingual resources. In *Proc. International Workshop on Paraphrasing*, Sapporo, Japan, July 2003.

3

Principal Direction Divisive Partitioning with Kernels and k-Means Steering

Dimitrios Zeimpekis and Efstratios Gallopoulos

Overview

Clustering is a fundamental task in data mining. We propose, implement, and evaluate several schemes that combine partitioning and hierarchical algorithms, specifically k-means and principal direction divisive partitioning (PDDP). Using available theory regarding the solution of the clustering indicator vector problem, we use 2-means to induce partitionings around fixed or varying cut-points. 2-means is applied either on the data or over its projection on a one-dimensional subspace. These techniques are also extended to the case of PDDP(l), a multiway clustering algorithm generalizing PDDP. To handle data that do not lend themselves to linear separability, the algebraic framework is established for a kernel variant, KPDDP. Extensive experiments demonstrate the performance of the above methods and suggest that it is advantageous to steer PDDP using k-means. It is also shown that KPDDP can provide results of superior quality than kernel k-means.

3.1 Introduction

Clustering is a major operation in text mining and in a myriad of other applications. We consider algorithms that assume the vector space representation for data objects [SB88], modeled as feature-object matrices (term-document matrices in text mining, hereafter abbreviated as tdms). Data collections are represented as $m \times n$ matrices A, where a_{ij} measures the importance of term i in document j. Two broad categories of clustering algorithms are partitional (the best known being k-means) and hierarchical, the latter being very desirable for Web-type applications [Ber06].

The k-means algorithm models clusters by means of their centroids. Starting from some initial guess for the k centroids, say $\{c_i\}_{i=1}^{k}$, representing the clusters, it iterates by first reclustering data depending on their distance from the current set of centroids, and then updating the centroids based on the new assignments. The progress of the algorithm can be evaluated using the objective function

$\sum_{i=1}^{k} \sum_{a_j \in \text{cluster}(i)} d(c_i, a_j)$, where d is some distance metric, e.g., quadratic Euclidean distance [Kog07]. This minimization problem for the general case is NP-hard. Two well-known disadvantages of the algorithm are that the generated clusters depend on the specific selection of initial centroids (e.g., random, in the absence of any data or application related information) and that the algorithm can be trapped at local minima of the objective function. Therefore, one run of k-means can easily lead to clusters that are not satisfactory and users are forced to initialize and run the algorithm multiple times.

On the other hand, a highly desirable feature, especially for Web-based applications, is the clustering structure imposed by hierarchical clustering algorithms (divisive or agglomerative), such as Single Link [SKK00]. Bisecting k-means is a divisive hierarchical variant of k-means that constructs binary hierarchies of clusters. The algorithm starts from a single cluster containing all data points, and at each step selects a cluster and partitions it into two subclusters using the classical k-means for the special case of $k = 2$. Unfortunately, the weaknesses of k-means (cluster quality dependent upon initialization and trappings at local minima) remain.

One particularly powerful class of clustering algorithms, with origins in graph partitioning ([DH73]), utilizes spectral information from the underlying tdm.[1] These algorithms provide the opportunity for the deployment of computational technologies from numerical linear algebra, an area that has seen enormous expansion in recent decades. In this paper we focus on principal direction divisive partitioning (PDDP), a clustering algorithm from this class that can be interpreted using principal component analysis (PCA). The algorithm first proposed by D. Boley [Bol98] is the primary source. PDDP can be quite effective when clustering text collections as it exploits sparse matrix technology and iterative algorithms for very large eigenvalue problems. See [Bol01, LB06, SBBG02, ZG03] as well as the recent monograph [Kog07] and references therein that present several variants of PDDP. At each iteration, PDDP selects one of the current clusters and partitions it into two subclusters using information from the PCA of the corresponding tdm. In fact, the basic PDDP algorithm can be viewed as a special case of a more general algorithm, PDDP(l), that constructs 2^l-ary cluster trees. PDDP is known to be an effective clustering method for text mining, where tdms are very large and extremely sparse. The algorithm requires the repeated extraction of leading singular vector information from the tdm[2] and submatrices thereof via the singular value decomposition (SVD). Even though, in general, the SVD is an expensive computation, it has long been known that significant savings can result when seeking only selected singular triplets from very sparse tdms [Ber92]. PDDP, however, is typically more expensive than k-means. Furthermore, the partitioning performed at each step of PDDP is duly determined by the outcome of the partial SVD of the cluster tdm (Section 3.2). Despite the convenient

---

[1] See Chris Ding's collection of tutorials and references at https://crd.lbl.gov/~cding/Spectral.

[2] More precisely, from the matrix resulting from the tdm after centering it by subtracting from each column its centroid.

deterministic nature of this step, it is easy to construct examples where PDDP produces inferior partitionings than k-means.

For some time now, we have been studying the characteristics of PDDP and have been considering ways to improve its performance. An early contribution in this direction was PDDP(l) ([ZG03]), a generalization along the lines of early work in multiway spectral graph partitioning [AKY99, AY95, HL95]. Our first original contribution here is to show how to leverage the power of k-means and some interesting recent theory ([DH04, ZHD$^+$01]) in order to better steer the partitioning decision at each iteration of PDDP. Our results confirm that such an approach leads to better performance compared to standard PDDP, in which the decision criteria are readily available from the spectral characteristics of the tdm.

Our second contribution is that we combine PDDP with kernel techniques, making the algorithm suitable for datasets not easily linearly separable. Specifically, we establish the necessary algebraic framework and propose KPDDP, a kernel version of PDDP as well as variants that we analyze and demonstrate that they return results of high quality compared to kernel k-means, albeit at higher cost.

The chapter is organized as follows. Section 3.2 describes algorithms that generalize PDDP and introduces the concept of k-means steering. Section 3.3 reviews kernel PCA, and Section 3.4 discusses the kernel versions of the above algorithms. Finally, Section 3.5 contains extensive experimental results.

As is frequently done, we will be using capital letters to denote matrices, lower-case letters for (column) vectors, and lower-case Greek letters for scalars. When referring to specific matrix elements (scalars), we will use the lower case Greek letter corresponding to the Latin capital letter. When referring to vectors (rows or columns) of a matrix, we will use the corresponding lower case Latin letter. We also use diag($\cdot$) to denote the diagonal of its argument.

3.2 PDDP Multipartitioning and Steering

It has long been known that the effectiveness of clustering algorithms depends on the application and data at hand. We focus on PDDP as an example of a spectral clustering algorithm that has been reported to be quite effective for high dimensional data, such as text, where standard density-based algorithms, such as DBSCAN, are not as effective because of their high computational cost and difficulty to achieve effective discrimination [BGRS99, SEKX98]. PDDP, on the other hand, extracts spectral information from the data and uses it to construct clusterings that can be of better quality than those provided by k-means. PDDP can be quite efficient in comparison to other agglomerative hierarchical algorithms and can be used either as an independent clustering tool or as a "preconditioning tool" for the initialization of k-means.

We next highlight some features of PDDP that motivate the proposals and variants of our chapter. At each step of PDDP, a cluster is selected and then partitioned in two subclusters using information from the PCA of the matrix corresponding to the data points belonging to the cluster. In this manner, the algorithm constructs a binary tree of clusters. More precisely, let p indicate the cluster node selected for partitioning at step i_p, $A^{(p)}$ the corresponding matrix (i.e., a submatrix of A with $n^{(p)}$ columns), $c^{(p)}$ its column centroid, and $C^{(p)} = (A^{(p)} - c^{(p)}e^\top)(A^{(p)} - c^{(p)}e^\top)^\top$ the corresponding covariance matrix. During the split process, ordinary PDDP uses the first principal component of $A^{(p)}$. Equivalently, if $[u^{(p_i)}, \sigma^{(p_i)}, v^{(p_i)}]$ is the ith singular triplet of the centered matrix $B^{(p)} = (A^{(p)} - c^{(p)}e^\top)$, the two subclusters are determined by the sign of the projection coefficients of each centered data point into the first principal component, $(u^{(p_1)})^\top(a^{(p_j)} - c^{(p)}) = \sigma^{(p_1)}v^{(p_{1j})}$, i.e., the sign of the corresponding element of $v^{(p_1)}$, since $\sigma^{(p_1)} > 0$.

Geometrically, the points $(u^{(p_1)})^\top(z - c^{(p)}) = 0$ define a hyperplane passing through the origin and perpendicular to $u^{(p_1)}$. Each data element is classified to one of two hyperplane-defined half-spaces. This can be extended to an 2^l-way partitioning strategy, based on information readily available from $l \geq 1$ singular vectors [ZG03]. This method, named PDDP(l), classifies data points to one of 2^l orthants, based on the specific sign combination of the vector $(U_l^{(p)})^\top(a^{(p_j)} - c^{(p)}) = \Sigma_l^{(p)}(V_l^{(p)}(j,:))^\top$, where $U_l^{(p)}\Sigma_l^{(p)}(V_l^p)^\top$ is the l-factor truncated SVD of $B^{(p)}$. Each one of $(u^{(p_i)})^\top(z - c^{(p)}) = 0$ defines a hyperplane passing through the origin orthogonal to $u^{(p_i)}$. In the sequel it would be useful to note that PDDP(l) reduces to the original PDDP algorithm when $l = 1$. Moreover, l is the dimension of the space in which lie the projection coefficients used to decide upon the splitting at each step of the algorithm. Finally, since PDDP(l) immediately classifies data into 2^l clusters, it can be used to partition $k = 2^l$ clusters in one single step.

Note that throughout our discussion, the cluster selected for partitioning will be the one with maximum scatter value, defined as $s_p = \|A^{(p)} - c^{(p)}e^\top\|_F^2$. This quantity is a measure of coherence, and it is the same as the one used by k-means [KMN⁺02, Llo82]. For convenience, we tabulate PDDP(l) in Algorithm 3.2.1.

Algorithm 3.2.1 Algorithm PDDP(l).

Input: Feature×object matrix A (tdm), desired number
    of clusters k, number of principal components l
Output: Structure of pddp_tree, with
$$\left\lceil \frac{k-1}{2^l-1} \right\rceil (2^l - 1) + 1 \text{ leaves}$$
Initialize pddp_tree(A);
for $i = 1 : \left\lceil \frac{k-1}{2^l-1} \right\rceil$
    Select a cluster node p to split with $n^{(p)}$ elements;
    Let $A^{(p)}$ be the tdm of cluster node p, $c^{(p)}$ the corresponding centroid;
    Compute the leading l singular triplets $[u^{(p_j)}, s^{(p_j)}, v^{(p_j)}]$,
        $j = 1, ..., l$ of $(A^{(p)} - c^{(p)}e^\top)$;
    Assign each column $a^{(p_j)}$ of $A^{(p)}$ to node $2 + (i - 2)2^l + j$,
        $j = 0 : 2^l - 1$ according to the signs of row j of $[v^{(p_1)}, ..., v^{(p_l)}]$;
    Update pddp_tree;
end

We have seen that with some work, PDDP classifies data from the cluster under consideration (possibly the entire dataset) into 2^l subclusters at the end of the first iteration. As outlined above, data are classified based on their orthant address, obtained from the corresponding sign combination of the l right singular vectors of the tdm. As we will see, there are cases where the sign combination is not a good cluster predictor and alternative classification techniques must be sought.

k-Means Steered PDDP

We next show how k-means can provide an effective steering mechanism for the partitioning at each step of PDDP(l). As we will see there are several alternatives; we first describe them in the context of the simplest version of PDDP. There, cluster membership of each datapoint is based on the sign of the corresponding element of the leading ($l = 1$) right singular vector of the centered tdm. Restating slightly, membership is decided based on whether the aforementioned element of the singular vector is smaller or larger than some "cut-point," which in this case is selected to be zero.

We call the "cluster indicator vector problem" the computation of an indicator vector d with elements $\delta_i \in \{\pm 1\}$, whose values could imply a specific cluster assignment, for example, assigning data element i to cluster C_1 (resp. C_2) if $\delta_i = 1$ (resp. "-1"). We make use of the following result:

Theorem 1 ([DH04]). *Let $A \in \mathbb{R}^{m \times n}$ be the tdm. For k-means clustering where $k = 2$, the continuous solution of the cluster indicator vector problem is the leading principal component, $r = [\rho_1, ..., \rho_n]^\top$, of A, that is, data will be assigned to each of the two clusters according to the following index partitioning: $C_1 = \{i | \rho_i \leq 0\}, C_2 = \{i | \rho_i > 0\}$, where $i = 1, ..., n$.*

The above theorem suggests that if we relax the condition that $\delta_i \in \{\pm 1\}$ and allow the elements of the indicator vector to take any real value, the resulting vector would be r. Therefore, one partitioning technique could be to classify every data element in the cluster that is a candidate for splitting based on the position of the corresponding element of r relative to 0. Theorem 1 suggests that splitting the data vectors according to r could be a good initialization for bisecting k-means. This technique can also be viewed as a refinement of the basic PDDP, in which 2-means is applied to ameliorate the initial splitting. We refer to this approach as PDDP-2MEANS.

Initial centroids of 2-means are given by the ordering of r, that is,

$$c_1 = (1/n_1) \sum_{i:\rho_i \leq 0} a_i \text{ and } c_2 = (1/n_2) \sum_{i:\rho_i > 0} a_i,$$

where n_1, n_2 denote respectively the numbers of negative or zero and positive elements of r. Therefore, the nondeterminism in (bisecting) k-means is eliminated.

We next note that even though r gives the solution of the relaxed problem, zero might not be the best cut-point. Indeed, Theorem 1 provides only an ordering for the data vectors, but not necessarily the best solution for the relaxed problem. In particular, another cut-point could lead to a smaller value for the k-means objective function. Specifically, we can check all possible splittings using the ordering implied by r, and select the one that optimizes the k-means objective function.

Lemma 1. *Let $A \in \mathbb{R}^{m \times n}$ and let $(j_1, j_2, \ldots, j_n)$ be an ordering of its columns. Then the optimal cut-point for 2-means can be obtained at cost $O(mn)$.*

Proof. Let $\{a_{j_1} a_{j_2} \ldots a_{j_n}\}$ the ordering of A's columns and $C_1 = \{a_{j_1} \ldots a_{j_l}\}$, $C_2 = \{a_{j_{l+1}} \ldots a_{j_n}\}$ a partition. The objective function value of C_1, C_2 is

$$s_1 = \sum_{i=j_1}^{j_l} \|a_i - c_1\|^2 = \sum_{i=j_1}^{j_l} \|a_i\|^2 - l\|c_1\|^2$$

$$s_2 = \sum_{i=j_{l+1}}^{j_n} \|a_i - c_2\|^2 = \sum_{i=j_{l+1}}^{j_n} \|a_i\|^2 - (n-l)\|c_2\|^2$$

while the objective function value for the clustering is given by $s = s_1 + s_2$. Consider now the partition $C_1 = \{a_{j_1} \ldots a_{j_l} a_{j_{l+1}}\}$, $C_2 = \{a_{j_{l+2}} \ldots a_{j_n}\}$. The new centroid vectors are given by

$$\hat{c}_1 = \frac{lc_1 + a_{j_{l+1}}}{l+1}, \hat{c}_2 = \frac{lc_2 - a_{j_{l+1}}}{n-l-1}$$

while

$$\hat{s} = \sum_{i=j_1}^{j_n} \|a_i\|^2 - (l+1)\hat{c}_1 - (n-l-1)\hat{c}_2.$$

Clearly, the update of the centroid vectors requires $O(m)$ operations as well as the computation of the new value of the objective function. This operation is applied $n - 1$ times and the proof follows.

Based on the above proof, we can construct an algorithm to determine the optimal cut-point corresponding to the elements of the leading principal component of A. The resulting splitting can be used directly in PDDP; alternatively, it provides a starting point for the application of 2-means. We refer to the algorithms corresponding to these two options as PDDP-OC and PDDP-OC-2MEANS respectively. The above techniques can be extended to the case of PDDP(l).

We next note that in the course of their 2-means phase, all the above enhancements of PDDP (PDDP-2MEANS, PDDP-OC, PDDP-OC-2MEANS) necessitate further operations between m-dimensional data vectors. Despite the savings implied by Lemma 1, costs can quickly become prohibitive for datasets of very high-dimensionality. On the other extreme, costs would be minimal for datasets consisting of only one feature. We can thus consider constructing algorithms in which we apply the above enhancements, not on the original dataset but on some compressed representation. We know, for example, that in terms of the Euclidean norm, the optimal unit rank approximation of the centered tdm can be written as $\sigma_1 u_1 v_1^\top$, where $\{\sigma_1, u_1, v_1\}$ is its largest singular triplet. Vector v_1 contains the coefficients of the projection of the centered tdm on the space spanned by u_1 and can be used as a one-dimensional encoding of the dataset. We further refine the usual PDDP policy, which utilizes zero as the cut-point, that is, employs a sign-based splitting induced by this vector. In this manner, however, the magnitude of the corresponding elements plays no role in the decision process.

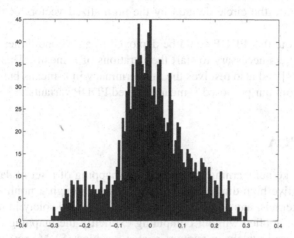

Fig. 3.1. Histogram of the projection coefficients employed by PDDP.

To see that this could lead to an inferior partitioning, we depict in Fig. 3.1 a histogram of the elements of v_1 corresponding to part of the Reuters-21578 collection. Many elements cluster around the origin; however, a value near -0.1 appears to be a more reasonable choice for cut-point than 0. Our next proposal is an algorithm that addresses this issue in the spirit of the previous enhancements of PDDP. Specifically, we propose the application of the 2-means algorithm on the elements of v_1. A key feature of this proposal is the fact that 2-means is applied on a one-dimensional dataset. We can thus evade NP-hardness, in fact we can compute the optimal cut-point so that the centroids of the two subsets of data points on either side of this point maximize the objective function of 2-means. To accomplish this, it is sufficient to sort the elements of v_1 and examine all possible partitions. Since the objective function for each attained partitioning can be computed in constant time and there are only $n - 1$ possible partitions, the entire process can be completed at a cost of only $O(n)$ operations. Note that this is a direct extension of Lemma 1 for the case of one-dimensional datasets.

We can apply the same principle to PDDP(l), where $l > 1$, by identifying the optimal cut-point along each one of the l leading principal components (equivalently, the leading right singular vectors). If $z := [\zeta_1 \ \ldots \ \zeta_l]^\top$ consists of the l cut-points that are optimal for each dimension, we can apply the basic PDDP(l) procedure by first shifting the origin by z. We denote by PDDP-OCPC the implied technique. Note that all the above policies eliminate the random start from 2-means and make it deterministic.

It is fair to mention here that the potential for alternative partitioning strategies was already mentioned in the original PDDP paper.[3] Two related approaches are [Dhi01] and [KDN04]. The former paper proposes the use of k-means to best assign scalar data into k clusters, the case $k = 2$ being the most prominent. In the latter, an algorithm called spherical PDDP (sPDDP for short) is proposed, employing $l = 2$ principal components during the split procedure and computing the optimal separation line into the circle defined by the normalized vectors of the projection coefficients.

We finally note that PDDP could be deployed as a preconditioner, to determine the k initial clusters necessary to start the iterations of k-means. This idea was described in [XZ04] and also resolves the indeterminacy in k-means but is fundamentally different from our proposed k-means steered PDDP variants.

3.3 Kernel PCA

The key idea in kernel learning methods is the mapping of a set of data points into a more informative high-dimensional feature space through a nonlinear mapping. Using Mercer kernels, any algorithm that can be expressed solely in terms of inner products, can be applied without computing explicitly the mapping. Kernel methods have been used widely in support vector machine (SVM) research [CST00]. In the context of clustering, Finley and Joachims in [FJ05] and Ben-Hur et al. in [BHHSV01] propose some SVM based clustering algorithms. Schölkopf et al. in [SSM98] and Camastra and Verri in [CV05] propose the kernel k-means algorithm, while Zhang and Chen in [ZC03] propose a kernel fuzzy c-means variant. Dhillon et al. in [DGK04] provide a useful connection between kernel k-means and spectral clustering.

PCA is an important statistical tool used in a wide range of applications where there is a need for dimensionality reduction. Kernel PCA (KPCA) is an extension of PCA that is applied to a vector space defined by a nonlinear mapping and has been used with success in a wide range of applications (e.g., [KS05, MMR+01, SM97, SSM98, YAK00]). In particular, let ϕ be a general mapping, $\phi : X \rightarrow F$, mapping input vectors $x \in X$ to vectors $\phi(x) \in F$. We assume that X is a subset of $\mathbb{R}^m$. KPCA amounts to the application of standard PCA to a new vector space, called feature space. Let $\langle \cdot, \cdot \rangle$ denote the Euclidean inner product. Inner products in F can be computed via the "kernel trick," that is, by defining an appropriate symmetric kernel function, $k(x, y)$ over $X \times X$, which satisfies Mercer's theorem ([CST00]) so that

---

[3] As stated in [Bol98], "... the values of the singular vector are used to determine the splitting ... in the simplest version of the algorithm we split the documents strictly according to the sign ..."

$$k(x^{(i)}, x^{(j)}) := \langle \phi(x^{(i)}), \phi(x^{(j)}) \rangle \tag{3.1}$$

for $x^{(i)}, x^{(j)} \in X$. There are various kernel functions, with the polynomial, $k(x, y) = (x^\top y + 1)^d$, and Gaussian, $k(x, y) = \exp(-\|x - y\|^2/\sigma^2)$, being two of the most common. It then becomes possible to express algorithms over F that use only inner products via kernels, specifically without ever computing the $\phi(x^{(i)})$'s.

Assume for now that the $\phi(x^{(i)})$'s are already centered. The covariance matrix of $\tilde{A} = [\phi(x^{(1)}) \ \ldots \ \phi(x^{(n)})]$ is

$$\tilde{C} = \sum_{i=1}^{n} \phi(x^{(i)})\phi(x^{(i)})^\top = \tilde{A}\tilde{A}^\top.$$

The principal components of $\tilde{A}$ are the eigenvectors of $\tilde{C}$, that is, the solutions of $\tilde{C}v = \lambda v$. Each v can be expressed as a linear combination of the columns of $\tilde{A}$, i.e., $v = \sum_{i=1}^{n} \psi_i \phi(x^{(i)})$. The latter can be rewritten as $v = \tilde{A}y$, where $y = [\psi_1, ..., \psi_n]^\top$; therefore, projecting the eigenvalue problem into each column of $\tilde{A}$, it turns out that the PCA of $\tilde{A}$ amounts to an eigenvalue problem for the Gram matrix, namely solutions of $Ky = \lambda y$ where $(K)_{i,j} := \kappa_{ij} = \langle \phi(x^{(i)}), \phi(x^{(j)}) \rangle$. This can be solved using Eq. (3.1) without forming the $\phi(x^{(i)})$'s, assuming that the Gram matrix K is available. The last step is the normalization of the eigenvectors $y^{(j)}, j = 1, \ldots, n$. This is necessary since $\langle v^{(j)}, v^{(j)} \rangle = 1$ must hold:

$$1 = \langle v^{(j)}, v^{(j)} \rangle = \langle \sum_{i=1}^{n} \psi_i^{(j)} \phi(x^{(i)}), \sum_{r=1}^{n} \psi_r^{(j)} \phi(x^{(r)}) \rangle$$

$$= \sum_{i,r=1}^{n} y_i^{(j)} \psi_r^{(j)} \langle \phi(x^{(i)}), \phi(x^{(r)}) \rangle$$

$$= \langle y^{(j)}, Ky^{(j)} \rangle = \lambda_j \langle y^{(j)}, y^{(j)} \rangle = \lambda_j \|y^{(j)}\|_2^2.$$

So, the final step is the normalization of $y^{(j)}$ according to

$$\|y^{(j)}\|_2 = \frac{1}{\sqrt{\lambda_j}}. \tag{3.2}$$

The coefficients of the projection of a column $\phi(x^{(r)})$ of $\tilde{A}$ into an eigenvector $v^{(j)}$ of $\tilde{C}$ are given by

$$\langle v^{(j)}, \phi(x^{(r)}) \rangle = \sum_{i=1}^{n} y_i^{(j)} \langle \phi(x^{(i)}), \phi(x^{(r)}) \rangle = (y^{(j)})^\top K_{:,r}$$

where $K_{:,r}$ denotes the rth column of the Gram matrix.

Setting Y_k to be the matrix formed by the k leading eigenvectors of K, $Y_k^\top K$ contains a k-dimensional representation for each vector $\phi(x^{(j)})$. If the $\phi(x^{(i)})$'s are not centered, it was shown in [SSM98] that the principal components of $\tilde{C}$ are given by the eigenvectors of

$$\tilde{K} = K - \frac{1}{n}ee^\top K - \frac{1}{n}Kee^\top + \frac{1}{n^2}ee^\top Kee^\top. \tag{3.3}$$

After some algebraic manipulation, $\tilde{K}$ can be expressed as $(I - P)K(I - P)$, where $P := ee^\top/n$ is an orthogonal projector. In this case the projection coefficients of the data points into the first k principal components are $Y_k^\top \tilde{K}$, where Y_k consists of the leading k eigenvectors of $\tilde{K}$.

3.4 Kernel Clustering

Our proposal is to use KPCA in place of PCA during the splitting step of PDDP. Assuming that, based on scatter, p is the cluster node selected for partitioning at step i_p, we accomplish this using information from the KPCA of the corresponding matrix $A^{(p)}$, that is, from the linear PCA of $\hat{B}^{(p)} = \Phi(B^{(p)}) = \Phi(A^{(p)} - c^{(p)}e^\top)$, where $\Phi(\cdot)$ denotes the matrix $\Phi(X)^{(j)} = \phi(x^{(j)})$ and $\phi(\cdot)$ a nonlinear mapping (e.g., polynomial kernel).

As indicated in Eq. (3.3), the principal components of $\hat{B}^{(p)}$ are given by the eigenvectors of

$$\hat{K}^{(p)} = K^{(p)} - MK^{(p)} - K^{(p)}M + MK^{(p)}M \tag{3.4}$$

where $M = \frac{1}{n^{(p)}}ee^\top$ and $[K_{ij}^{(p)}] = [\langle\phi(a^{(p_i)} - c^{(p)}), \phi(a^{(p_j)} - c^{(p)})\rangle]$, the Gram matrix corresponding to cluster p. The projection coefficients of $\hat{B}^{(p)}$ into the top l eigenvectors of the covariance matrix $(\hat{B}^{(p)})(\hat{B}^{(p)})^\top$ are given by

$$(V_l^{(p)})^\top \hat{B}^{(p)} \in \mathbb{R}^{l \times n^{(p)}} \tag{3.5}$$

where $V_l^{(p)}$ are the l leading eigenvectors of $[K_{ij}^{(p)}]$. Note that the normalization condition for the eigenvectors of $\hat{K}^{(p)}$ also causes the normalization of $V^{(p)}$'s columns according to Eq. (3.2).

Algorithm 3.4.1 Algorithm KPDDP(l).

Input: Feature×object matrix A (tdm), desired number
 of clusters k, number of principal components l
Output: Structure of `pddp_tree`, with
 $\left\lceil \frac{k-1}{2^l-1} \right\rceil (2^l - 1) + 1$ leaves
Initialize `pddp_tree`(A);
for $i = 1 : \left\lceil \frac{k-1}{2^l-1} \right\rceil$
 Select a cluster node p to split according to (3.6); Let $A^{(p)}$ be the tdm
  of cluster p, $c^{(p)}$ the corresponding centroid;
 Form the Gram matrix $K^{(p)}$;
 Compute the leading l eigenvectors and eigenvalues of $\hat{K}^{(p)}$;
 Normalize the eigenvectors $V^{(p_i)}$, $i = 1, \ldots, l$ of $K^{(p)}$ according to (3.2);
 Assign each column $a^{(p_j)}$ of $A^{(p)}$ to cluster node $2 + (i - 2)2^l + j$,
  $j = 0 : 2^l - 1$ according to the signs of row j of $(V_l^{(p)})^\top \hat{B}^{(p)}$;
 Update `pddp_tree`;
end

Regarding the cluster selection step, note that

$$s_{p_\phi} = \|\Phi(A^{(p)}) - \hat{c}^{(p)}(e^{(p)})^\top\|_F^2$$

$$= \sum_{j=1}^{n^{(p)}} \|\hat{a}^{(p_j)} - \hat{c}^{(p)}\|_F^2 = \sum_{j=1}^{n^{(p)}} \|\hat{a}^{(p_j)} - \hat{c}^{(p)}\|_2^2$$

where $\hat{c}^{(p)}$ the centroid of $\Phi(A^{(p)})$ into the feature space defined by $\phi(\cdot)$. Since $\|x - y\|_2^2 = \langle x, x\rangle^2 + \langle y, y\rangle^2 - 2\langle x, y\rangle$, it follows that the scatter value can be computed from

$$s_{p_\phi} = \sum_{j=1}^{n^{(p)}} \|\phi(a^{(p_j)})\|_2^2 + \sum_{j=1}^{n^{(p)}} \|\hat{c}^{(p)}\|_2^2 - 2\sum_{j=1}^{n^{(p)}} \langle \phi(a^{(p_j)}), \hat{c}^{(p)}\rangle$$

$$= \sum_{j=1}^{n^{(p)}} \langle \phi(a^{(p_j)}), \phi(a^{(p_j)})\rangle + n^{(p)}\langle \frac{\Phi(A^{(p)})e}{n^{(p)}}, \frac{\Phi(A^{(p)})e}{n^{(p)}}\rangle$$

$$-2\sum \langle \phi(a^{(p_j)}), \Phi(A^{(p)})e^{(p)}\rangle$$

$$= \sum_{j=1}^{n^{(p)}} \langle \phi(a^{(p_j)}), \phi(a^{(p_j)})\rangle + \frac{\|\bar{K}^{(p)}e\|_1}{n^{(p)}} - 2\frac{\|\bar{K}^{(p)}e\|_1}{n^{(p)}}$$

where $\bar{K}_{ij}^{(p)} = \langle \phi(a^{(p_i)}), \phi(a^{(p_j)})\rangle$ denotes the Gram matrix of $A^{(p)}$. Therefore,

$$s_{p_\phi} = \text{trace}(\bar{K}^{(p)}) - 2\frac{\|\bar{K}^{(p)}e\|_1}{n^{(p)}}. \tag{3.6}$$

We call the proposed algorithm KPDDP(l) (Kernel PDDP(l)). The algorithm is tabulated in Algorithm 3.4.1. As in the basic algorithm, at each step the cluster node p with the largest scatter value [Eq. (3.6)] is selected and partitioned into 2^l subclusters. Each member of the selected node is classified into the subcluster defined by the combination of its projection coefficients [Eq. (3.5)] into the l principal components of $\hat{B}^{(p)}$. In our implementation, we experimented with polynomial and Gaussian kernels. Our formulation, however, is general and can be applied for any kernel. The leading eigenpairs of the covariance matrix were computed from the SVD of the centered tdm using PROPACK [Lar]. This consists of a very efficient MATLAB implementation that applies Lanczos bidiagonalization with partial reorthogonalization to compute selected singular triplets. The algorithm requires access to a routine to form matrix-vector products $K^{(p)}x$, but does not necessitate the explicit construction of the Gram matrix $\hat{K}^{(p)}$.

Kernel k-Means

Kernel learning has already been applied in k-means clustering; e.g., [CV05, SSM98] outline efficient implementations. As in the linear case, each cluster is represented by its centroid into the new feature space, $\hat{c}^{(p)} = \frac{1}{n_p}(\phi(a_{p_1}) + \ldots + \phi(a_{p_n}))$; however, there is no need to compute $\hat{c}^{(p)}$'s. In particular, the algorithm operates iteratively as k-means by assigning each data point to the nearest cluster and updates centroids using the last assignment. The norm of the Euclidean distance of x, y equals to $\|x - y\|_2^2 = \langle x, x \rangle + \langle y, y \rangle - 2\langle x, y \rangle$. Denoting by K the Gram matrix of A and using MATLAB notation, the distance of a single data point a_j from a centroid $\hat{c}^{(p)}$ is $\langle a_j, a_j \rangle + \texttt{diag}(K)_{p_1 \ldots p_n} - 2K_{j,(p_1 \ldots p_n)}$. $\langle a_j, a_j \rangle$'s can be computed initially, and then during each iteration the cluster membership indication matrix can be computed as $P = ex^\top + ye^\top - 2Z$, where x contains the norms of A's columns, y the norms of centroids computed from a partial sum of K's elements once in each iteration, and Z the $k \times n$ matrix with elements $\zeta_{i,k} = \langle \hat{c}^{(i)}, a_j \rangle$. The algorithm assigns each element j to cluster $i_j = \arg\{\min_i P(i, j)\}$.

k-means can be extended as in the linear case in order to get a hierarchical clustering solution (kernel bisecting k-means). Furthermore, using the same reasoning, we can combine the kernel versions of PDDP and k-means in order to derive more effective clustering solutions. In particular, we propose the use of the kernel 2-means algorithm during the splitting process of KPDDP (we refer to it as KPDDP-2MEANS). Finally, we can extend the PDDP-OCPC variant in the kernel learning framework.[4]

3.5 Experimental Results

We next conduct extensive experiments in order to evaluate the impact of the proposed techniques. For this purpose, we use the well known Reuters-21578, Ohsumed

---

[4] We note that methods PDDP-OC and PDDP-OC-2MEANS can also be extended in kernel variants (see Theorem 3.5 in [DH04]); however, we observed that this results in a high computational overhead that makes these approaches prohibitive.

(part of the TREC filtering track), and CLASSIC3 (a merge of MEDLINE, CRAN-FIELD, and CISI) collections. For the linear case, we use the ModApte split of the Reuters-21578 collection as well as part of the Ohsumed collection composed of those documents that belong to a single cluster. Furthermore, we use CLASSIC3 for a toy example. For the nonlinear case, we use four datasets, named REUTj ($j = 1, ..., 4$), constructed from the ModApte split with varying number of clusters and cluster sizes.[5] Table 3.1 depicts the characteristics of those collections. TMG [ZG06] has been used for the construction of the tdms. Based on [ZG06], we used logarithmic local term and IDF global weightings with normalization, stemming, and stopword removal, removing also terms that appeared only once in the collection. Our experiments were conducted on a Pentium IV PC with 1-GB RAM using MATLAB.

Table 3.1. Dataset statistics

Feature	MODAPTE	OHSUMED	CLASSIC3	REUT1	REUT2	REUT3	REUT4
Documents	9,052	3,672	3,891	840	1,000	1,200	3,034
Terms	10,123	6,646	7,823	2,955	3,334	3,470	5,843
Terms/document	60	81	77	76	75	60	78
tdm nonzeros (%)	0.37	0.76	0.64	1.60	1.43	0.37	84
Number of clusters	52	63	3	21	10	6	25

In the following discussion, we denote by k the sought number of clusters. For each dataset we ran all algorithms for a range of k. In particular, denoting by r the true number of clusters for a dataset, we ran all algorithms for $k = 4 : 3 : k_{max}$ and $k = 8 : 7 : k_{max}$ for some $k_{max} > r$ in order to record the results of PDDP(l) and related variants for $l = 1, 2, 3$. For all k-means variants we have conducted 10 experiments with random initialization of centroids and recorded the minimum, maximum, and mean values of attained accuracy and run time. Although we present only mean-value results, minimum and maximum values are important to the discussion that follows. For the SVD and eigendecomposition we used the MATLAB interface of the PROPACK software package [Lar]. For the algorithms' evaluation, we use the objective function of k-means (and PDDP), the entropy and run-time measures.

Fig. 3.2 depicts the objective function, entropy values, and run time for all variants, for the linear case and datasets MODAPTE and OHSUMED. Although k-means appears to give the best results between all variants and all measures, we note that these plots report mean values attained by k-means and related variants. In practice, a single run of k-means may lead to poor results. As a result, a "good" partitioning may require several executions of the algorithm. Compared to the basic algorithm, its hierarchical counterpart (bisecting k-means) appears to degrade the quality of clustering and suffers from the same problems as k-means. On the other hand, PDDP appears to give results inferior to k-means. Regarding the proposed variants, we note that all techniques always improve PDDP and bisecting k-means in most cases, while

---

[5] We will call the ModApte and Ohsumed datasets as MODAPTE, OHSUMED.

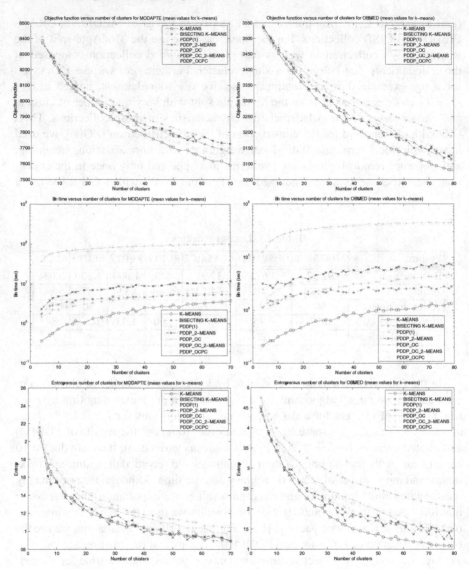

Fig. 3.2. Objective function, entropy values, and run time for k-means, PDDP, and variants.

approaching the clustering quality of k-means. PDDP-OC-2MEANS and PDDP-OC appear to give the best results; however, their high run times make them relatively expensive solutions. On the other hand, PDDP-2MEANS and PDDP-OCPC provide results similar to k-means without impacting significantly the overall efficiency of PDDP.

Fig. 3.3 (upper) depicts the quality of clustering of PDDP-OCPC vs. PDDP(l) for $l = 1, 2, 3$. PDDP-OCPC appears to improve significantly the efficiency of PDDP(l)

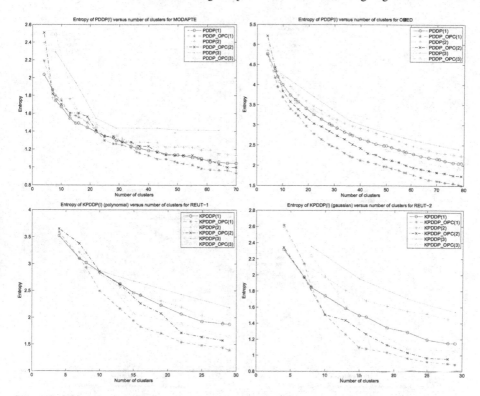

Fig. 3.3. Objective function and entropy values for PDDP and PDDP-OCPC and $l = 1, 2, 3$, linear (upper) and nonlinear (lower) case.

Table 3.2. Confusion matrices for CLASSIC3 for PDDP (upper left quadrant for $k = 3$ and right for $k = 4$) and PDDP-OCPC (lower left quadrant for $k = 3$ and right for $k = 4$)

Cluster	1	2	3	1	2	3	4
Class 1	12	6	1,015	12	1,015	4	2
Class 2	1,364	14	20	1,364	20	8	6
Class 3	2	1,392	66	2	66	788	604
Class 1	0	9	1,024	0	1,024	7	2
Class 2	1,253	29	116	1,253	116	23	6
Class 3	0	1,431	29	0	29	917	514

for equal values of parameter l. A simple example that demonstrates the success of PDDP-OCPC is given in Table 3.2, where we give the confusion matrices for PDDP and PDDP-OCPC for CLASSIC3 and $k = 3, 4$. PDDP-OCPC appears to approximate better the cluster structure of the collection by producing "cleaner" clusters.

In Figs. 3.4, 3.5 (polynomial and gaussian kernels, respectively) we give the results for the kernel versions of the algorithms for datasets REUT1-REUT4. As in the linear case, kernel k-means appears to give better results than kernel PDDP. However,

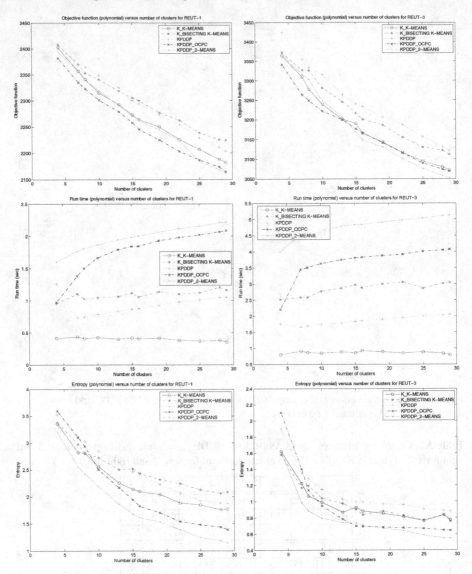

Fig. 3.4. Objective function, entropy values and runtime for kernel versions of k-means, PDDP and variants (polynomial kernels).

our hybrid schemes appear to improve even the k-means algorithm at low additional run time. Fig. 3.3 (lower) depicts the quality of clustering of the kernel versions PDDP-OCPC vs. PDDP(l) for $l = 1, 2, 3$. As in the linear case, PDDP-OCPC appears to provide significant improvements over PDDP. It is worth noting that in our experiments, results with the linear versions appeared to be uniformly better than results with the kernel implementations. This does not cause concern since our goal

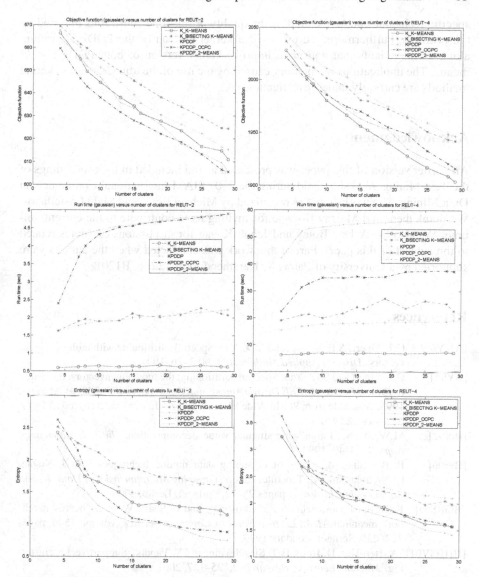

Fig. 3.5. Objective function, entropy values and runtime for kernel versions of k-means, PDDP and variants (gaussian kernel).

was to improve kernel k-means algorithm along deterministic approaches that are expected to give better results in case there are strong nonlinearities in the data at hand.

The above results indicate that our hybrid clustering methods that combine k-means and PDDP can be quite successful in addressing the nondeterminism in k-means, while achieving at least its "average-case" effectiveness. The selection of a specific technique can be dictated by the quality or run-time constraints imposed by the problem. Furthermore, we proposed a kernel version of the PDDP algorithm, along some variants that appear to improve kernel version of both PDDP and k-means. The implications on memory caused by the use of the tdm Gramian in kernel methods are currently under investigation.

Acknowledgments

An earlier version of this paper was presented at and included in the proceedings of the Text Mining Workshop held during the 2007 SIAM International Conference on Data Mining. The workshop was organized by Michael Berry and Malu Castellanos. We thank them and Murray Browne for inviting us to contribute to the current volume. We also thank Dan Boley and Jacob Kogan for discussions on topics related to the subject of this paper. Part of this work was conducted when the authors were supported by a University of Patras K. Karatheodori grant (no. B120).

References

[AKY99]   C.J. Alpert, A.B. Kahng, and S.-Z. Yao. Spectral partitioning with multiple eigenvectors. *Discrete Applied Mathematics*, 90:3–26, 1999.

[AY95]   C.J. Alpert and S.-Z. Yao. Spectral partitioning: the more eigenvectors, the better. In *Proc. 32nd ACM/IEEE Design Automation Conf.*, pages 195–200. ACM Press, 1995. Available from World Wide Web: http://doi.acm.org/10.1145/217474.217529.

[Ber92]   M.W. Berry. Large scale singular value decomposition. *Int'l. J. Supercomp. Appl.*, 6:13–49, 1992.

[Ber06]   P. Berkhin. A survey of clustering data mining techniques. In J. Kogan, C. Nicholas, and M. Teboulle, editors, *Grouping Multidimensional Data: Recent Advances in Clustering*, pages 25–72. Springer, Berlin, 2006.

[BGRS99]   K. Beyer, J. Goldstein, R. Ramakrishnan, and U. Shaft. When is "nearest neighbor" meaningful? In *Lecture Notes in Computer Science*, volume 1540, pages 217–235. Sprnger, London, 1999.

[BHHSV01]   A. Ben-Hur, D. Horn, H.T. Siegelmann, and V. Vapnik. Support vector clustering. *Machine Learning Research*, 2:125–137, 2001.

[Bol98]   D. Boley. Principal direction divisive partitioning. *Data Mining and Knowledge Discovery*, 2(4):325–344, 1998.

[Bol01]   D. Boley. A scalable hierarchical algorithm for unsupervised clustering. In R. Grossman, C. Kamath, P. Kegelmeyer, V. Kumar, and R. Namburu, editors, *Data Mining for Scientific and Engineering Applications*. Kluwer Academic Publishers, Norwell, MA, 2001.

[CST00]   N. Cristianini and J. Shawe-Taylor. *An Introduction to Support Vector Machines and Other Kernel-Base Learning Methods*. Cambridge University Press, Cambridge, UK, 2000.

[CV05]   F. Camastra and A. Verri. A novel kernel method for clustering. *IEEE Transactions on Pattern Analysis and Machine Intelligence*, 27(5):801–804, 2005.

[DGK04]   I.S. Dhillon, Y. Guan, and B. Kulis. Kernel k-means: spectral clustering and normalized cuts. In *Proc. 10th ACM SIGKDD*, pages 551–556, ACM Press, New York, 2004.

[DH73]   W.E. Donath and A.J. Hoffman. Lower bounds for the partitioning of graphs. *IBM J. Res. Develop.*, 17:420–425, 1973.

[DH04]   C. Ding and X. He. Cluster structure of k-means clustering via principal component analysis. In *PAKDD*, pages 414–418, 2004. Available from World Wide Web: http://springerlink.metapress.com/openurl.asp? genre=article&issn=0302-9743&volume=3056&spage=414.

[Dhi01]   I.S. Dhillon. Co-clustering documents and words using bipartite spectral graph partitioning. In *Proc. 7th ACM SIGKDD*, pages 269–274, ACM Press, New York, 2001.

[FJ05]   T. Finley and T. Joachims. Supervised clustering with support vector machines. In *ICML '05: Proceedings of the 22nd international conference on Machine learning*, pages 217–224, New York, 2005.

[HL95]   B. Hendrickson and R. Leland. An improved spectral graph partitioning algorithm for mapping parallel computations. *SIAM J. Sci. Comput.*, 16(2):452–469, 1995. Available from World Wide Web: citeseer.nj.nec.com/ hendrickson95improved.html.

[KDN04]   J. Kogan, I.S. Dhillon, and C. Nicholas. Feature selection and document clustering. In M. Berry, editor, *A Comprehensive Survey of Text Mining*. Springer, New York, 2004.

[KMN$^+$02] T. Kanungo, D.M. Mount, N.S. Netanyahu, D. Platko, and A.Y. Wu. An efficient k-means clustering algorithm: analysis and implementation. *IEEE Trans. PAMI*, 24(7):881–892, 2002.

[Kog07]   J. Kogan. *Introduction to Clustering Large and High-Dimensional Data*. Cambridge University Press, New York, 2007.

[KS05]   E. Kokiopoulou and Y. Saad. PCA and kernel PCA using polynomial filtering: a case study on face recognition. In *SIAM Conf. on Data Mining*, 2005.

[Lar]   R.M. Larsen. Propack: a software package for the symmetric eigenvalue problem and singular value problems on Lanczos and Lanczos bidiagonalization with partial reorthogonalization. Available from World Wide Web: http: //sun.stanford.edu/~rmunk/PROPACK/. Stanford University.

[LB06]   D. Littau and D. Boley. Clustering very large datasets with PDDP. In J. Kogan, C. Nicholas, and M. Teboulle, editors, *Grouping Multidimensional Data: Recent Advances in Clustering*, pages 99–126. Springer, New York, 2006.

[Llo82]   S.P. Lloyd. Least squares quantization in PCM. *IEEE Trans. Information Theory*, 28:129–137, 1982.

[MMR$^+$01] K.R. Müller, S. Mika, G. Rätsch, K. Tsuda, and B. Schölkopf. An introduction to kernel-based learning algorithms. *IEEE Transactions on Neural Networks*, 12(2):181–202, 2001.

[SB88]   G. Salton and C. Buckley. Term weighting approaches in automatic text retrieval. *Information Processing and Management*, 24(5):513–523, 1988.

[SBBG02]   S. Savaresi, D. Boley, S. Bittanti, and G. Gazzaniga. Choosing the cluster to split in bisecting divisive clustering algorithms. In *Second SIAM International Conference on Data Mining (SDM'2002)*, 2002.

[SEKX98]   J. Sander, M. Ester, H.-P. Kriegel, and X. Xu. Density-based clustering in spatial databases: the algorithm GDBSCAN and its applications. *Data Mining and Knowledge Discovery*, 2(2):169–194, 1998.

[SKK00]   M. Steinbach, G. Karypis, and V. Kumar. A comparison of document cluster-
           ing techniques. In *6th ACM SIGKDD, World Text Mining Conference*, Boston,
           MA, 2000. Available from World Wide Web: `citeseer.nj.nec.com/`
           `steinbach00comparison.html`.

[SM97]   A.J. Schölkopf, B. Smola and K.R. Müller. Kernel principal component analysis.
           In *Proc. International Conference on Artificial Neural Networks*, pages 583–588,
           1997.

[SSM98]   B. Schölkopf, A.J. Smola, and K.R. Müller. Nonlinear component analysis as a
           kernel eigenvalue problem. *Neural Computation*, 10(5):1299–1319, 1998.

[XZ04]   S. Xu and J. Zhang. A parallel hybrid Web document clustering algorithm and
           its performance study. *J. Supercomputing*, 30(2):117–131, 2004.

[YAK00]   M.H. Yang, N. Ahuja, and D.J. Kriegman. Face recognition using kernel eigen-
           faces. In *Proc. International Conference on Image Processing*, 2000.

[ZC03]   D.Q. Zhang and S.C. Chen. Clustering incomplete data using kernel-based fuzzy
           c-means algorithm. *Neural Processing Letters*, 18(3):155–162, 2003.

[ZG03]   D. Zeimpekis and E. Gallopoulos. PDDP(l): towards a flexible principal direction
           divisive partitioning clustering algorithm. In D. Boley, I. Dhillon, J. Ghosh, and
           J. Kogan, editors, *Proc. Workshop on Clustering Large Data Sets (held in con-
           junction with the Third IEEE Int'l. Conf. Data Min.)*, pages 26–35, Melbourne,
           FL, November 2003.

[ZG06]   D. Zeimpekis and E. Gallopoulos. TMG: A MATLAB toolbox for generat-
           ing term-document matrices from text collections. In J. Kogan, C. Nicholas,
           and M. Teboulle, editors, *Grouping Multidimensional Data: Recent Advances in
           Clustering*, pages 187–210. Springer, New York, 2006.

[ZHD$^+$01] H. Zha, X. He, C. Ding, M. Gu, and H. Simon. Spectral relaxation for k-
           means clustering. In *NIPS*, pages 1057–1064, 2001. Available from World
           Wide Web: `http://www-2.cs.cmu.edu/Groups/NIPS/NIPS2001/`
           `papers/psgz/AA41.ps.gz`.

4

Hybrid Clustering with Divergences

Jacob Kogan, Charles Nicholas, and Mike Wiacek

Overview

Clustering algorithms often require that the entire dataset be kept in the computer memory. When the dataset is large and does not fit into available memory, one has to compress the dataset to make the application of clustering algorithms possible. The balanced iterative reducing and clustering algorithm (BIRCH) is designed to operate under the assumption that "the amount of memory available is limited, whereas the dataset can be arbitrarily large" [ZRL97]. The algorithm generates a compact dataset summary minimizing the I/O cost involved. The summaries contain enough information to apply the well-known k-means clustering algorithm to the set of summaries and to generate partitions of the original dataset. (See [BFR98] for an application of quadratic batch k-means, and [KT06] for an application of k-means with Bregman distances to summaries.) An application of k-means requires an initial partition to be supplied as an input. To generate a "good" initial partition of the summaries, this chapter suggests a clustering algorithm, PDsDP, motivated by PDDP [Bol98]. We report preliminary numerical experiments involving sequential applications of BIRCH, PDsDP, and k-means/deterministic annealing to the Enron email dataset.

4.1 Introduction

Clustering very large datasets is a contemporary data mining challenge. A number of algorithms capable of handling large datasets that do not fit into memory have been reported in the literature (see [Ber06]). A sequence of algorithms such that the output of algorithm i is the input to algorithm $i + 1$ may be useful in this context (see [KNV03]). For example, a sequential application of clustering procedures that first compress data $\mathcal{A}$, then cluster the compressed data $\mathcal{B}$, and finally recover a partitioning of the original dataset $\mathcal{A}$ from the partitioning of the compressed dataset $\mathcal{B}$ has been reported, for example, in [BFR98, LB06, KNW07].

BIRCH [ZRL97] is one of the first clustering algorithms that computes summaries (or sufficient statistics), and uses summaries instead of the original dataset for clustering. An application of the classical batch k-means to summaries is reported in [BFR98]. The proposed algorithm clusters summaries, and, through association, partitions the original dataset without access to its elements.

The choice of a "good" initial partition for k-means is an additional clustering challenge. The principal direction divisive partitioning (PDDP) introduced by D. Boley [Bol98] can be used to address this task. PDDP substitutes for the original dataset $\mathcal{A}$ the one-dimensional "approximation" $\mathcal{A}'$, bisects $\mathcal{A}'$ into two clusters, and recovers the induced two-cluster partition of $\mathcal{A}$. The algorithm is then applied to each of these two clusters recursively. The subdivision is stopped when, for example, the desired number of clusters is generated for the dataset $\mathcal{A}$.

When applying PDDP to the summaries, $\mathcal{B}$ one should keep in mind that each summary represents a vector set, hence the approximation $\mathcal{B}'$ should reflect the size, the "quality," and the "geometry" of the vector set. In this chapter we present the principal directions divisive partitioning (PDsDP), a computationally efficient procedure to build the approximation set $\mathcal{B}'$. The algorithm applies the building block of PDDP to each summary separately.

The chapter is organized as follows. Section 4.2 introduces centroids and two specific families of distance-like functions. In Section 4.3 we briefly review the well-known clustering algorithms BIRCH, k-means, and DA. The principal directions divisive partitioning algorithm (which is the main contribution of the chapter) is introduced in detail. Section 4.4 reports on numerical experiments with the Enron email datasets. The particular distance-like function $d(\mathbf{x}, \mathbf{y})$ we use for numerical experiments is a combination of weighted squared Euclidean distance and Kullback–Leibler divergence. Section 4.5 indicates how the clustering strategy can be modified to reduce its computational complexity and better fit the data, and concludes the chapter.

4.2 Distance-Like Functions and Centroids

We start with some preliminary notations. For a finite vector set $\pi \subset \mathbf{R}^n$ and a distance-like function $d(\mathbf{x}, \mathbf{y})$, we define the centroid $\mathbf{c}(\pi)$ as a solution of the minimization problem:

$$\mathbf{c}(\pi) = \arg\min\left\{\sum_{\mathbf{a} \in \pi} d(\mathbf{x}, \mathbf{a}), \ \mathbf{x} \in \mathcal{C}\right\}, \qquad (4.1)$$

where $\mathcal{C}$ is a specified subset of $\mathbf{R}^n$ (as, for example, $\mathbf{R}_+^n$, or an $n - 1$ dimensional sphere $\mathcal{S} = \{\mathbf{x} : \|\mathbf{x}\| = 1\}$). The quality of π is defined by

$$Q(\pi) = \sum_{\mathbf{a} \in \pi} d(\mathbf{c}(\pi), \mathbf{a}). \qquad (4.2)$$

For a dataset $\mathcal{A} = \{\mathbf{a}_1, \ldots, \mathbf{a}_m\} \subset \mathbf{R}^n$ the quality of partition $\Pi = \{\pi_1, \ldots, \pi_k\}$, $\pi_i \cap \pi_j = \emptyset$ if $i \neq j$, $\pi_1 \cup \cdots \cup \pi_k = \mathcal{A}$ is given by

$$Q(\Pi) = \sum_{i=1}^{k} Q(\pi_i).\qquad(4.3)$$

The degree of difficulty involved in solving (4.1) depends on the function $d(\cdot, \cdot)$, and the set $\mathcal{C}$. In this chapter we shall be concerned with two specific families of distance-like functions: the Bregman and Csiszár divergences.

4.2.1 Bregman Divergence

Let $\psi : \mathbf{R}^n \rightarrow (-\infty, +\infty]$ be a closed proper convex function. Suppose that ψ is continuously differentiable on int(dom ψ) $\neq \emptyset$. The Bregman divergence D_ψ : dom $\psi \times$ int(dom ψ) $\rightarrow \mathbf{R}_+$ is defined by

$$D_\psi(\mathbf{x}, \mathbf{y}) = \psi(\mathbf{x}) - \psi(\mathbf{y}) - \nabla\psi(\mathbf{y})(\mathbf{x} - \mathbf{y})\qquad(4.4)$$

where $\nabla\psi$ is the gradient of ψ. This function measures the convexity of ψ, that is,

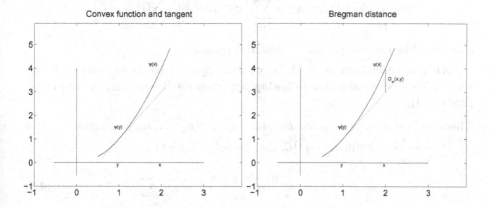

Fig. 4.1. Bregman divergence.

$D_\psi(\mathbf{x}, \mathbf{y}) \geq 0$ if and only if the gradient inequality for ψ holds, that is, if and only if ψ is convex. With ψ strictly convex one has $D_\psi(\mathbf{x}, \mathbf{y}) \geq 0$ and $D_\psi(\mathbf{x}, \mathbf{y}) = 0$ iff $\mathbf{x} = \mathbf{y}$ (see Figure 4.1).

Note that $D_\psi(\mathbf{x}, \mathbf{y})$ is not a distance (it is, in general, not symmetric and does not satisfy the triangle inequality). The well-known examples of Bregman divergences are:

1. The squared Euclidean distance $D_\psi(\mathbf{x}, \mathbf{y}) = \|\mathbf{x} - \mathbf{y}\|^2$ (with $\psi(\mathbf{x}) = \|\mathbf{x}\|^2$), and
2. The Kullback–Leibler divergence

$$D_\psi(\mathbf{x}, \mathbf{y}) = \sum_{j=1}^{n} \mathbf{x}[j] \log \frac{\mathbf{x}[j]}{\mathbf{y}[j]} + \mathbf{y}[j] - \mathbf{x}[j]$$

$$\left(\text{with } \psi(\mathbf{x}) = \sum_{j=1}^{n} \mathbf{x}[j] \log \mathbf{x}[j] - \mathbf{x}[j]\right).$$

For additional interesting examples we refer the reader to [BMDG04], [TBD$^+$06]. Bregman divergences are convex with respect to the first argument, hence centroids (solutions for minimization problem (4.1)) can be computed efficiently. Since the function $D_\psi(\mathbf{x}, \mathbf{y})$ is not symmetric, by reversing the order of variables in D_ψ, that is,

$$\overleftarrow{D_\psi}(\mathbf{x}, \mathbf{y}) = D_\psi(\mathbf{y}, \mathbf{x}) = \psi(\mathbf{y}) - \psi(\mathbf{x}) - \nabla\psi(\mathbf{x})(\mathbf{y} - \mathbf{x}), \qquad (4.5)$$

one obtains a different distance-like function (compare with (4.4)). For example, using the kernel

$$\psi(\mathbf{x}) = \sum_{j=1}^{n} \mathbf{x}[j] \log \mathbf{x}[j] - \mathbf{x}[j], \qquad (4.6)$$

we obtain

$$\overleftarrow{D_\psi}(\mathbf{x}, \mathbf{y}) = D_\psi(\mathbf{y}, \mathbf{x}) = \sum_{j=1}^{n} \left[\mathbf{y}[j] \log \frac{\mathbf{y}[j]}{\mathbf{x}[j]} + \mathbf{x}[j] - \mathbf{y}[j]\right], \qquad (4.7)$$

a distance-like function popular in clustering research.

While in general distances $\overleftarrow{D_\psi}(\mathbf{x}, \mathbf{y})$ given by (4.5) fail to be convex with respect to the first variable, the following surprising result was recently reported in [BMDG04].

Theorem 1. *If $d(\mathbf{x}, \mathbf{y})$ is given by (4.5), $\pi = \{\mathbf{a}_1, \ldots, \mathbf{a}_m\} \subset$ dom ψ, and $\dfrac{\mathbf{a}_1 + \cdots + \mathbf{a}_m}{m} \in int(\text{dom } \psi) \bigcap \mathcal{C}$, then the solution for (4.1) is*

$$\mathbf{c}(\pi) = \mathbf{c}(\pi) = \frac{\mathbf{a}_1 + \cdots + \mathbf{a}_m}{m} = \frac{1}{|\pi|} \sum_{\mathbf{a} \in \pi} \mathbf{a}, \qquad (4.8)$$

where $|\pi|$ is the size of π.

Switching the variables has the potential to change the result. Indeed, for $d(\mathbf{x}, \mathbf{y}) = \overleftarrow{D_\psi}(\mathbf{x}, \mathbf{y}) = D_\psi(\mathbf{y}, \mathbf{x}) = \sum_{j=1}^{n} \left[\mathbf{y}[j] \log \dfrac{\mathbf{y}[j]}{\mathbf{x}[j]} + \mathbf{x}[j] - \mathbf{y}[j]\right]$, the centroid is given by

the *arithmetic* mean. On the other hand, if $d(\mathbf{x}, \mathbf{y}) = D_\psi(\mathbf{x}, \mathbf{y}) = \sum_{j=1}^{n} \mathbf{x}[j] \log \dfrac{\mathbf{x}[j]}{\mathbf{y}[j]} + \mathbf{y}[j] - \mathbf{x}[j]$, then the centroid is given by the *geometric* mean (see [TBD$^+$06]).

4.2.2 Csiszár Divergence

Let $\Phi = \{\varphi : \mathbf{R} \to (-\infty, +\infty]\}$ be the class of functions satisfying (1)–(4) below. We assume that for each $\varphi \in \Phi$ one has dom $\varphi \subseteq [0, +\infty)$, $\varphi(t) = +\infty$ when $t < 0$ and φ satisfies the following:

1. φ is twice continuously differentiable on int(dom φ) = $(0, +\infty)$.
2. φ is strictly convex on its domain.
3. $\lim_{t \to 0^+} \varphi'(t) = -\infty$.
4. $\varphi(1) = \varphi'(1) = 0$ and $\varphi''(1) > 0$.

Given $\varphi \in \Phi$, for $\mathbf{x}, \mathbf{y} \in \mathbf{R}^n$ we define $d_\varphi(\mathbf{x}, \mathbf{y})$ by

$$d_\varphi(\mathbf{x}, \mathbf{y}) = \sum_{j=1}^{n} \mathbf{y}[j] \varphi \left(\frac{\mathbf{x}[j]}{\mathbf{y}[j]} \right). \tag{4.9}$$

The function $d_\varphi(\mathbf{x}, \mathbf{y})$ is convex with respect to $\mathbf{x}$ and with respect to $\mathbf{y}$.[1] Recall that centroid computations require us to solve a minimization problem (4.1) involving d_φ. Assumptions (1) and (2) above ensure existence of global minimizers, and assumption (3) enforces the minimizer to stay in the positive octant. Condition (4) is a normalization that allows for the handling of vectors in $\mathbf{R}_+^n$ (rather than probabilities).

The functional d_φ enjoys basic properties of a distance-like function, namely $\forall (\mathbf{x}, \mathbf{y}) \in \mathbf{R}^n \times \mathbf{R}^n$. One has:

$$d_\varphi(\mathbf{x}, \mathbf{y}) \geq 0 \quad \text{and} \quad d_\varphi(\mathbf{x}, \mathbf{y}) = 0 \quad \text{iff} \quad \mathbf{x} = \mathbf{y}.$$

Indeed, the strict convexity of φ and Assumption 4 above imply

$$\varphi(t) \geq 0, \quad \text{and} \quad \varphi(t) = 0 \text{ iff } t = 1.$$

The choice $\varphi(t) = -\log t + t - 1$, with dom$\varphi = (0, +\infty)$ leads to

$$d_\varphi(\mathbf{x}, \mathbf{y}) \equiv KL(\mathbf{y}, \mathbf{x}) = \sum_{j=1}^{n} \left[\mathbf{y}[j] \log \frac{\mathbf{y}[j]}{\mathbf{x}[j]} + \mathbf{x}[j] - \mathbf{y}[j] \right]. \tag{4.10}$$

The above examples show that the functions

$$d(\mathbf{x}, \mathbf{y}) = \|\mathbf{x} - \mathbf{y}\|^2, \text{ and } d(\mathbf{x}, \mathbf{y}) = \sum_{j=1}^{n} \left[\mathbf{y}[j] \log \frac{\mathbf{y}[j]}{\mathbf{x}[j]} + \mathbf{x}[j] - \mathbf{y}[j] \right]$$

are Bregman divergences convex with respect to both variables. In the numerical experiments section we work with the Bregman divergence

$$d(\mathbf{x}, \mathbf{y}) = \frac{\nu}{2} \|\mathbf{x} - \mathbf{y}\|^2 + \mu \sum_{j=1}^{n} \left[\mathbf{y}[j] \log \frac{\mathbf{y}[j]}{\mathbf{x}[j]} + \mathbf{x}[j] - \mathbf{y}[j] \right], \ \nu \geq 0, \ \mu \geq 0.$$

$$\tag{4.11}$$

In the next section we review briefly two well known clustering algorithms, and introduce the PDsDP algorithm that mimics PDDP [Bol98].

---

[1] In [Ros98] and [WS03] the function $d(\mathbf{x}, \mathbf{y})$ is *required* to be convex with respect to $\mathbf{x}$. Csiszár divergences are natural candidates for distance-like functions satisfying this requirement.

4.3 Clustering Algorithms

We start with a brief review of the BIRCH algorithm.

4.3.1 BIRCH

Let $\Pi = \{\pi_1, \ldots, \pi_M\}$ be a partition of $\mathcal{A} = \{\mathbf{a}_1, \ldots, \mathbf{a}_m\}$. For $i = 1, \ldots, M$ we denote by

1. $\mathbf{b}_i$ the centroid $\mathbf{c}(\pi_i)$ of π_i,
2. m_i or $m(\mathbf{b}_i)$ the size of π_i,
3. q_i the quality $Q(\pi_i)$ of the cluster π_i.

For proof of the next result, consult [TBD$^+$06]:

$$Q(\Pi) = \sum_{i=1}^{k} Q(\pi_i) + \sum_{i=1}^{k} m_i d(\mathbf{c}, \mathbf{b}_i) = \sum_{i=1}^{k} Q(\pi_i) + \sum_{i=1}^{k} m_i \left[\psi(\mathbf{b}_i) - \psi(\mathbf{c})\right]$$

(4.12)

where

$$\mathbf{c} = \mathbf{c}(\mathcal{A}) = \frac{m_1}{m}\mathbf{b}_1 + \cdots + \frac{m_k}{m}\mathbf{b}_k, \text{ and } m = m_1 + \cdots + m_k.$$

Formula (4.12) paves the way to approach the clustering of $\mathcal{A}$ along the two lines:

- Given a positive real constant R (which controls the "spread" of a cluster), an integer L (which controls the size of a cluster), p already available clusters $\pi_1, \ldots \pi_p$ (i.e., p summaries $(m_i, q_i, \mathbf{b}_i)$), and a vector $\mathbf{a} \in \mathbf{R}^n$, one can compute $Q(\pi_i \cup \{\mathbf{a}\})$, $i = 1, \ldots, p$ using $\mathbf{a}$ and summaries only. If for some index i

$$Q(\pi_i \cup \{\mathbf{a}\}) = q_i + d\left(\frac{\mathbf{a}+m_i\mathbf{b}_i}{1+m_i}, \mathbf{a}\right) + m_i d\left(\frac{\mathbf{a}+m_i\mathbf{b}_i}{1+m_i}, \mathbf{b}_i\right) < R$$

and

$$m_i + 1 \leq L,$$

(4.13)

then $\mathbf{a}$ is assigned to π_i, and the triplet $(m_i, q_i, \mathbf{b}_i)$ is updated. Otherwise $\{\mathbf{a}\}$ becomes a new cluster π_{p+1} (this is the basic BIRCH construction).

An alternative "greedy" and computationally more expensive strategy is to assign $\mathbf{a}$ to the cluster π that satisfies (4.13) and minimizes "average quality" $\dfrac{Q(\pi_i \cup \{\mathbf{a}\})}{m_i + 1}$ over all clusters π_i satisfying (4.13). In the numerical experiments presented in Section 4.4 we follow the "greedy" approach.

- Once a partition $\Pi = \{\pi_1, \ldots, \pi_M\}$ of $\mathcal{A}$ is available, one can cluster the set $\mathcal{B} = \{\mathbf{b}_1, \ldots, \mathbf{b}_M\}$. Note that the M cluster partition $\{\pi_1, \ldots, \pi_M\}$ of $\mathcal{A}$ associates each subset $\pi^{\mathcal{B}} \subseteq \mathcal{B}$ with a subset $\pi^{\mathcal{A}} \subseteq \mathcal{A}$ through

$$\pi^{\mathcal{A}} = \bigcup_{\mathbf{b}_j \in \pi^{\mathcal{B}}} \pi_j.$$

Hence a k cluster partition $\Pi_B = \{\pi_1^B, \ldots, \pi_k^B\}$ of the set B is associated with a k cluster partition $\Pi_A = \{\pi_1^A, \ldots, \pi_k^A\}$ of the set A through

$$\pi_i^A = \bigcup_{b_j \in \pi_i^B} \pi_j, \ i = 1, \ldots, k. \tag{4.14}$$

One can, therefore, apply k-means to the smaller dataset B to generate a partition of the dataset A (this approach is suggested in [BFR98] for batch k-means equipped with $d(\mathbf{x}, \mathbf{y}) = \|\mathbf{x} - \mathbf{y}\|^2$).

Consider a k cluster partition $\Pi_B = \{\pi_1^B, \ldots, \pi_k^B\}$ of the set B and the associated k cluster partition $\Pi_A = \{\pi_1^A, \ldots, \pi_k^A\}$ of the set A. Consider, for example, $\pi_1^B = \{\mathbf{b}_1, \ldots, \mathbf{b}_p\}$ with $\mathbf{c}\left(\pi_1^B\right)$ and the corresponding cluster $\pi_1^A = \pi_1 \cup \cdots \cup \pi_p$. Due to (4.12) one has

$$Q\left(\pi_1^A\right) = \sum_{j=1}^{p} Q(\pi_j) + \sum_{b \in \pi_1^B} m(\mathbf{b}) d\left(\mathbf{c}\left(\pi_1^B\right), \mathbf{b}\right).$$

Repetition of this argument for other clusters π_i^B and summing the corresponding expressions leads to

$$\sum_{i=1}^{k} Q\left(\pi_i^A\right) = \sum_{l=1}^{M} Q(\pi_l) + \sum_{i=1}^{k} \sum_{b \in \pi_i^B} m(\mathbf{b}) d\left(\mathbf{c}\left(\pi_i^B\right), \mathbf{b}\right). \tag{4.15}$$

We set $Q_B\left(\Pi_B\right) = \sum_{i=1}^{k} \sum_{b \in \pi_i^B} m(\mathbf{b}) d\left(\mathbf{c}\left(\pi_i^B\right), \mathbf{b}\right)$, note that $\sum_{l=1}^{M} Q(\pi_l) = Q(\Pi)$ is a constant, and arrive at the following formula

$$Q\left(\Pi_A\right) = Q(\Pi) + Q_B\left(\Pi_B\right). \tag{4.16}$$

We next describe PDsDP–a deterministic algorithm that generates partitions of the set B (these partitions will be fed to a k-means–type algorithm at the next step of the clustering procedure).

4.3.2 PDsDP

The PDDP algorithm approximates the dataset A by the projection of the set on the line l that provides the best least squares approximation for A (see [Bol98]). The line is defined by the arithmetic mean $\mathbf{c}$ of A, and the principal eigenvector of the matrix

$$\mathbf{M} = \sum_{a \in A} (\mathbf{a}_i - \mathbf{c}) (\mathbf{a}_i - \mathbf{c})^T. \tag{4.17}$$

An application of BIRCH to the original dataset A generates a partition $\{\pi_1, \ldots, \pi_M\}$ and only the sufficient statistics $(|\pi_i|, Q(\pi_i), \mathbf{b}_i)$, $i = 1, \ldots, k$ are available (see Figure 4.2). To apply PDDP to the set $\{\mathbf{b}_1, \ldots, \mathbf{b}_M\}$, one has to take onto account

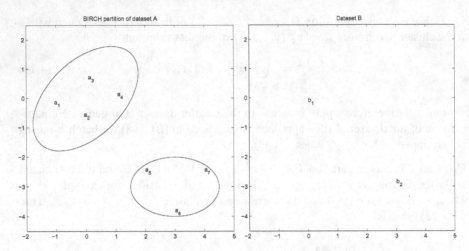

Fig. 4.2. BIRCH 2 cluster partition of dataset $\mathcal{A}$, and the summaries $\mathcal{B}$.

each cluster size $|\pi_i|$ as well as the spatial distribution of the cluster's elements. Figure 4.3 shows that the triplets $(|\pi_i|, Q(\pi_i), \mathbf{b}_i)$ alone do not capture the information required to generate the best least squares approximation line l for the set $\mathcal{A}$.

Indeed, consider the following two different datasets:

$$\mathcal{A}_1 = \left\{ \begin{bmatrix} 1 \\ 0 \end{bmatrix}, \begin{bmatrix} -1 \\ 0 \end{bmatrix}, \begin{bmatrix} 2 \\ 0 \end{bmatrix} \right\}, \text{ and } \mathcal{A}_2 = \left\{ \begin{bmatrix} \frac{1}{\sqrt{2}} \\ \frac{1}{\sqrt{2}} \end{bmatrix}, \begin{bmatrix} -\frac{1}{\sqrt{2}} \\ -\frac{1}{\sqrt{2}} \end{bmatrix}, \begin{bmatrix} 2 \\ 0 \end{bmatrix} \right\}.$$

While the corresponding best least squares approximation lines l are different (see Figure 4.3), the sufficient statistics $(|\pi_i|, Q(\pi_i), \mathbf{b}_i)$, $i = 1, 2$ for the BIRCH-generated partitions $\Pi_1 = \{\pi_1^1, \pi_2^1\}$ and $\Pi_2 = \{\pi_1^2, \pi_2^2\}$ (see Table 4.1) are identi-

Table 4.1. Different datasets with identical sufficient statistics

π_1^1	π_2^1	π_1^2	π_2^2
$\left\{ \begin{bmatrix} 1 \\ 0 \end{bmatrix}, \begin{bmatrix} -1 \\ 0 \end{bmatrix} \right\}$	$\left\{ \begin{bmatrix} 2 \\ 0 \end{bmatrix} \right\}$	$\left\{ \begin{bmatrix} \frac{1}{\sqrt{2}} \\ \frac{1}{\sqrt{2}} \end{bmatrix}, \begin{bmatrix} -\frac{1}{\sqrt{2}} \\ -\frac{1}{\sqrt{2}} \end{bmatrix} \right\}$	$\left\{ \begin{bmatrix} 2 \\ 0 \end{bmatrix} \right\}$

cal. They are

$$\left(2, 2, \begin{bmatrix} 0 \\ 0 \end{bmatrix} \right) \text{ and } \left(1, 0, \begin{bmatrix} 2 \\ 0 \end{bmatrix} \right).$$

Reconstruction of the best least squares approximation lines l based on triplets $(|\pi_i|, Q(\pi_i), \mathbf{b}_i)$ alone is therefore not possible.

To build the principal direction, most of the iterative algorithms compute the product $\mathbf{Mx}$. Formula (4.17) indicates that access to the entire dataset $\mathcal{A} =$

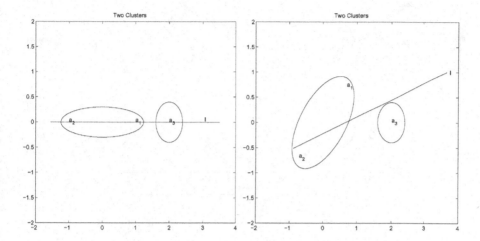

Fig. 4.3. Two BIRCH-generated clusters with identical sufficient statistics and different lines l.

$\{\mathbf{a}_1, \ldots, \mathbf{a}_m\}$ is needed to determine the projection line l. When the dataset $\mathcal{A}$ does not fit into memory, we propose to approximate $\mathcal{A}$ by the dataset $\mathcal{A}' = \{\mathbf{a}_1', \ldots, \mathbf{a}_m'\}$ and to use the covariance matrix

$$\mathbf{M}' = \sum_{i=1}^{m} (\mathbf{a}_i' - \mathbf{c}')(\mathbf{a}_i' - \mathbf{c}')^T. \tag{4.18}$$

While the size of $\mathcal{A}'$ is identical to the size of $\mathcal{A}$, to compute the product $\mathbf{M}'\mathbf{x}$, in addition to the triplets $(|\pi_i|, Q(\pi_i), \mathbf{c}(\pi_i))$, one needs to know only M additional vectors $\mathbf{x}_{1i}$ and scalars λ_{1i} (see Theorem 2).

The suggested approximations is motivated by PDDP. For each $i = 1, \ldots, M$, we propose to identify the principal eigenvector $\mathbf{x}_{1i}$ and the corresponding largest eigenvalue λ_{1i} of the covariance matrix corresponding to the cluster π_i:

$$\mathbf{M}_i = \sum_{\mathbf{a} \in \pi_i}^{|\pi_i|} (\mathbf{a} - \mathbf{b}_i)(\mathbf{a} - \mathbf{b}_i)^T \tag{4.19}$$

(see Figure 4.4, top left). For $\mathbf{a} \in \pi$ the vector $\mathbf{a}'$ is defined as the orthogonal projection of $\mathbf{a}$ on the line $\mathbf{l}_i = \{\mathbf{b}_i + \mathbf{x}_{1i}t, -\infty < t < \infty\}$ defined by the mean $\mathbf{b}_i$ and the cluster's principal direction vector $\mathbf{x}_{1i}$ (see Figure 4.4, bottom center).

Theorem 2. *The matrix vector product* $\mathbf{M}'\mathbf{x} = \left[\sum_{\mathbf{a}' \in \mathcal{A}'} (\mathbf{a}' - \mathbf{c})(\mathbf{a}' - \mathbf{c})^T\right] \mathbf{x}$ *is given by*

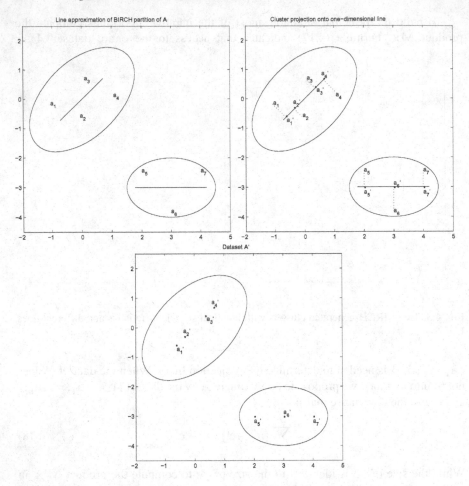

Fig. 4.4. Principal directions (top left), projections (top right), and "linear" approximation $\mathcal{A}'$ (bottom center).

$$\mathbf{M}'\mathbf{x} = \sum_{i=1}^{M}\mathbf{b}_i m_i \left(\mathbf{b}_i^T \mathbf{x}\right) + \sum_{i=1}^{M}\mathbf{x}_{1i}\lambda_{1i}\left(\mathbf{x}_{1i}^T\mathbf{x}\right) - m\mathbf{c}\left(\mathbf{c}^T\mathbf{x}\right).$$

Hence, in addition to sufficient statistics $(|\pi_i|, Q(\pi_i), \mathbf{b}_i)$, $i = 1, \ldots, M$ *required by BIRCH, computation of* $\mathbf{M}'\mathbf{x}$ *requires the eigenvectors* $\mathbf{x}_{1i}$ *along with the eigenvalues* λ_{1i}, $i = 1, \ldots, M$.

Proof. For $\mathbf{a} \in \pi_i$ the vector $\mathbf{a}'$ is the projection of $\mathbf{a}$ on the line defined by a unit norm vector $\mathbf{x}_i$ and the centroid $\mathbf{b}_i$ of the cluster π_i, that is,

$$\mathbf{a}' = \mathbf{b}_i + \mathbf{x}_i t(\mathbf{a}), \ t(\mathbf{a}) = \mathbf{x}_i^T (\mathbf{a} - \mathbf{b}_i)$$

with

(4.20)

$$\sum_{\mathbf{a} \in \pi_i} t(\mathbf{a}) = 0, \text{ and } \sum_{\mathbf{a} \in \pi_i} t^2(\mathbf{a}) = \lambda_i$$

(when π_i contains m_i identical copies of $\mathbf{b}_i$ we set $\mathbf{x}_i = 0$, and $\lambda_i = 0$). A straightforward substitution completes the proof.

We note that in information retrieval (IR) applications with data vectors $\mathbf{a} \in \mathbf{R}_+^n$, the vector $\mathbf{x}_i$ is as sparse as $\mathbf{b}_i$. Hence when high-quality BIRCH-generated clusters containing documents with many words in common are available, the centroids $\mathbf{b}_i$ as well as principal direction vectors $\mathbf{x}_{1i}$ are sparse.

Let $\lambda_{1i}, \lambda_{2i}, \ldots, \lambda_{Ni}$ and $\mathbf{x}_{1i}, \mathbf{x}_{2i}, \ldots, \mathbf{x}_{Ni}$ be the N largest eigenvalues of the cluster π_i covariance matrix along with the corresponding eigenvectors. Let $\mathbf{a}'$ be an orthogonal projection of $\mathbf{a} \in \pi_i$ on the affine subspace passing through $\mathbf{b}_i$ and spanned by the N eigenvectors $\mathbf{x}_{1i}, \mathbf{x}_{2i}, \ldots, \mathbf{x}_{Ni}$. A straightforward computation immediately leads to the following generalization of Theorem 2.

Theorem 3. *The matrix vector product* $\mathbf{M}'\mathbf{x} = \left[\sum_{\mathbf{a}' \in \mathcal{A}'} (\mathbf{a}' - \mathbf{c})(\mathbf{a}' - \mathbf{c})^T \right] \mathbf{x}$ *is given by*

$$\mathbf{M}'\mathbf{x} = \sum_{i=1}^{M} \mathbf{b}_i m_i \left(\mathbf{b}_i^T \mathbf{x} \right) + \sum_{j=1}^{N} \left[\sum_{i=1}^{M} \mathbf{x}_{ji} \lambda_{ji} \left(\mathbf{x}_{ji}^T \mathbf{x} \right) \right] - m\mathbf{c} \left(\mathbf{c}^T \mathbf{x} \right).$$

As N increases the quality of the approximation $\mathbf{a}'$ improves. Additional memory is needed to accommodate N eigenvectors and N eigenvalues for each BIRCH-generated cluster π_i, $i = 1, \ldots, M$. Provided the centroids $\mathbf{b}_i$ are sparse, and N is not too large, this memory requirement may be negligible.

While this chapter focuses on high-dimensional IR applications, when the vector space dimension n is relatively small, Theorem 3 opens the way to computation of the dataset $\mathcal{A}$ principal direction when only summaries $(|\pi_i|, Q(\pi_i), \mathbf{b}_i)$ along with n eigenvectors and eigenvalues are available for each BIRCH-generated cluster π_i.

To indicate the number of eigenvectors N used to generate the approximations $\mathbf{a}'$, we shall refer to the algorithm as PDs$_N$DP. Numerical results reported in Section 4.4 are generated by PDs$_1$DP and PDs$_2$DP.

4.3.3 k-Means

First we briefly recall batch and incremental k-means (see [DHS01]).

The batch k–means algorithm is an iterative two-step procedure. An iteration of the algorithm transforms a partition Π into a partition Π' so that $Q(\Pi) \geq Q(\Pi') \geq 0$. The iterations run as long as $Q(\Pi) - Q(\Pi')$ exceeds a user specified $\mathtt{tol} \geq 0$.

The two-step procedure that builds Π' from $\Pi = \{\pi_1, \ldots, \pi_k\}$ is briefly outlined next.

1. Use (4.1) to compute centroids $c\left(\pi_i\right)$, $i = 1, \ldots, k$.
2. For a set of k "centroids" $\{c_1, \ldots, c_k\}$, build the updated partition $\Pi' = \{\pi'_1, \ldots, \pi'_k\}$:

$$\pi'_i = \{a : a \in \mathcal{A} \; d\left(c_i, a\right) \leq d\left(c_l, a\right) \text{ for each } l = 1, \ldots, k\} \tag{4.21}$$

(we break ties arbitrarily). Note that, in general, $c\left(\pi_i\right) \neq c_i\left(\pi'_i\right)$.

This popular clustering algorithm was outlined already in the 1956 work of Steinhaus [Ste56]. While fast and simple to implement, this two-step procedure ignores the difference between the "old" and "new" centroids, and often "misses" better quality partitions (for a discussion see [TBD$^+$06]).

An iteration of incremental k-means examines partitions Π' generated from Π by removing a single vector a from cluster π_i and assigning it to cluster π_j. If the reassignment leads to a better partition, then the obtained partition becomes the iteration outcome. Otherwise a different reassignment is examined. The stopping criterion for this algorithm is the same as for the batch k–means algorithm.

While each iteration of incremental k-means changes the cluster affiliation of a single vector only, the iteration is based on the exact computation of $Q(\Pi)$ and $Q(\Pi')$. An application of a sequence of batch k-means iterations followed by a single incremental k-means iteration combines the speed of the former and the accuracy of the later. The computational cost of the additional incremental step depends on the specific distance-like function and may come virtually for free (see [DKN03], [TBD$^+$06]). We shall refer to the merger of the batch and incremental k-means as just the k-means clustering algorithm. The "merger" algorithm was recently suggested in [ZKH99], [HN01], and [Kog01]. Since the number of different partitions of a finite set $\mathcal{A}$ is finite, one can safely run k-means with $\mathtt{tol} = 0$ (and we select $\mathtt{tol} = 0$ for numerical experiments described in Section 4.4).

An application of quadratic batch k-means to summaries generated by BIRCH is reported in [BFR98], and k-means equipped with Bregman divergences and applied to summaries is reported in [KT06].

A different approach to clustering transforms k-means into a finite dimensional optimization problem, and opens the way to application of available optimization algorithms.

4.3.4 smoka

Rather than focusing on partitions, one can search for the set of k "centroids" $\{x_1, \ldots, x_k\} \subset \mathbf{R}^n$, or the vector $x = (x_1, \ldots, x_k) \in \mathbf{R}^{nk} = \mathbf{R}^N$. Since the "contribution" of the data vector a_i to the partition cost is $\min_{1 \leq l \leq k} d\left(x_l, a_i\right)$, we consider the objective function $F(x) = \sum_{i=1}^{m} \min_{1 \leq l \leq k} d\left(x_l, a_i\right)$. The clustering problem is now reduced to a continuous optimization problem in a finite dimensional Euclidean space $\mathbf{R}^N$:

$$\min_{x \in \mathbf{R}^N} F(x). \tag{4.22}$$

This is a *nonconvex* and *nonsmooth* optimization problem. Approximation of F by a family of smooth functions F_s as a way to handle (4.22) was suggested recently in [TK05] specifically for $d(\mathbf{x}, \mathbf{a}) = \|\mathbf{x} - \mathbf{a}\|^2$, and in [Teb07] for general distance-like functions (mathematical techniques dealing with smooth approximations can be traced to the classic manuscript [HLP34]). The particular choice of the family $F_s(\mathbf{x})$ given by

$$F_s(\mathbf{x}) = \sum_{i=1}^{m} -s \log \left(\sum_{l=1}^{k} e^{-\frac{d(\mathbf{x}_l, \mathbf{a}_i)}{s}} \right) \qquad (4.23)$$

leads to a special case of a simple iterative algorithm smoothed k-means algorithm (smoka) that generates a sequence of centroids $\mathbf{x}(t)$ with

$$F_s(\mathbf{x}(t)) \geq F_s(\mathbf{x}(t+1)) \geq -sm \log k.$$

The iterations run until $F_s(\mathbf{x}(t)) - F_s(\mathbf{x}(t+1)) \geq \texttt{tol}$, where $\texttt{tol} > 0$ is user specified [TK05].

Due to the lack of convexity, smoka convergence to a global minimum is not guaranteed (in general smoka converges to a critical point only). The computational effort of one iteration of smoka does not exceed that of k-means [Teb07] (hence in Section 4.4 we report and compare the number of iterations).

In 1990 Rose, Gurewitz, and Fox [RGF90] introduced a remarkable clustering algorithm called deterministic annealing (DA). The algorithm focuses on centroids rather than on partitions. The algorithm is inspired by principles of statistical mechanics. DA is used in order to avoid local minima of the given nonconvex objective function (see [RGF90]). Simulated annealing (SA, see [KGV83]) is a well-known method for solving nonconvex optimization problems. This stochastic method is motivated by its analogy to statistical physics and is based on the Metropolis algorithm [MRR+53]. Unlike simulated annealing, the DA algorithm replaces the stochastic search by a deterministic search, whereby the annealing is performed by minimizing a cost function (the free energy) directly, rather than via stochastic simulations of the system dynamics. Building on this procedure, global minima is obtained by minimizing the cost function (the free energy) while gradually reducing a parameter that plays the role of "temperature" (smoothing parameter s of smoka). The "right" rate of temperature change should lead to the global minimum of the objective function.

Among other things DA offers "the ability to avoid many poor local optima" (see [Ros98], p. 2210). More recently [Ros98], the DA has been restated within a purely probabilistic framework within basic information theory principles. The annealing process consists of keeping a system at equilibrium given by the minimum of the free energy while gradually reducing the temperature, so that the ground state is achieved at the limit of a low temperature. Many reports in the literature have indicated that the DA algorithm outperforms the standard k-means clustering algorithm, and it can be also used in the context of other important applications (see [Ros98] and references therein). The "right" rate of temperature decrease still remains an open problem.

The clustering algorithm smoka equipped with the quadratic Euclidean distance coincides with the deterministic annealing with *fixed* temperature. The numerical results collected in Section 4.4 provide insight into the performance of smoka vs. k-means. Several additional contributions consider k-means as a finite dimensional optimization problem (see [NK95], [ZHD99], [Kar99], [Yu05]).

Note that while clustering summaries $\mathcal{B} = \{b_1, \dots, b_M\}$, one has to modify $F(x)$ and $F_s(x)$ as follows:

$$F(\mathbf{x}) = \sum_{i=1}^{M} m_i \min_l d\left(\mathbf{x}_l, \mathbf{b}_i\right), F_s(\mathbf{x}) = -s\sum_{i=1}^{M} m_i \log \left(\sum_{l=1}^{k} e^{-\frac{d(\mathbf{x}_l, \mathbf{b}_i)}{s}}\right). \quad (4.24)$$

Implementation of an algorithm minimizing the modified $F_s(x)$ is identical to smoka.

4.4 Numerical Experiments

In this section we report on numerical experiments with the Enron email dataset.[2] The dataset contains 517,431 email messages located in 3501 subdirectories. There are 14 large (over 300 kb) files that apparently contain attachments.[3] We remove these large files from the collection, and process the remaining 517417 files.

In all the experiments 5000 "best" terms are selected (see [TBD+06] for the term selection procedure) to represent the documents in the vector space model (see [BB99]), and the tfn is applied to normalize the vectors (see [CK99]). In what follows we refer to a clustering algorithm equipped with the distance-like function (4.11) as "(ν, μ) clustering algorithm."

The three-step clustering procedure includes:

1. (ν, μ) BIRCH (to generate a set of summaries),
2. PDs$_1$DP or PDs$_2$DP (to generate an initial partition for k-means/smoka),
3. (ν, μ) k-means/smoka (to partition the summaries $\mathcal{B}$ and, by association, the original dataset $\mathcal{A}$).

In experiments reported below, k-means, applied to a finite set of partitions, runs with tol = 0, while smoka, minimizing functions of a vector variable, runs with tol = 0.0001, and $s = 0.0001$. The choice of s is motivated by the bound

$$0 \leq F(\mathbf{x}) - F_s(\mathbf{x}) \leq sm \log k$$

where m is the dataset size, and k is the desired number of clusters [Teb07]. We run the experiments with two extreme weights $(2, 0)$, $(0, 1)$, and the mixed weight $(50, 1)$ (this weight is selected to equate contributions of quadratic and logarithmic parts of $Q(\mathcal{A})$).

Since each email message contains a time stamp, in an attempt to simulate a data stream and to improve BIRCH clustering, we sort the messages with respect to time and feed them in this order to BIRCH.

---

[2] Available at http://www.cs.cmu.edu/~enron/
[3] For example \maildir\dorland-c\deleted_items\20

The upper bound for BIRCH generated cluster size is $L = 100$, and the upper bound for cluster average quality $\dfrac{Q(\pi)}{|\pi|}$ as a fraction of $\dfrac{Q(\mathcal{A})}{|\mathcal{A}|}$, the maximal cluster average quality, is reported in Table 4.2 for the three choices of parameters (ν, μ) along with additional characteristics of the collections $\mathcal{A}$ and $\mathcal{B}$.

Table 4.2. Collection: Enron email dataset; size, upper bound for cluster average quality, average cluster size, and sparsity for the original dataset $\mathcal{A}$, and (ν, μ) BIRCH generated vector sets $\mathcal{B}$

Dataset/(ν, μ)	$\mathcal{A}$	$\mathcal{B}/(2,0)$	$\mathcal{B}/(0,1)$	$\mathcal{B}/(50,1)$
Size	517,417	6073	6185	5322
Max clus. av. quality	na	0.9	0.8	0.9
Av. clus. size	na	85.19	83.65	97.22
Sparsity	1.5%	25%	25%	28%

We compute sparsity of a vectors set $\{\mathbf{a}_1, \ldots, \mathbf{a}_p\} \in \mathbf{R}^n$ as $\dfrac{1}{n \cdot p} \displaystyle\sum_{i=1}^{p} n(\mathbf{a}_i)$,

where $n(\mathbf{a})$ is the number of nonzero entries of a vector $\mathbf{a}$.

In these experiments PDs_NDP, $N = 1, 2$ generate 10 clusters. Results of sequential applications of $(2, 0)$ BIRCH, $\text{PDs}_1\text{DP}/\text{PDs}_2\text{DP}$, and $(2, 0)$ k-means/ smoka are reported in Table 4.3. The results collected in Table 4.3 show that two

Table 4.3. Collection: Enron email dataset; number of iterations and quality Q of $517, 417$ vector dataset $\mathcal{A}$ partition generated by (2,0) BIRCH, $\text{PDs}_1\text{DP}/\text{PDs}_2\text{DP}$, (2,0) k-means, and (2,0) smoka; the size of the dataset $\mathcal{B}$ is 6073, the vector space dimension is 5000, total of 10 clusters

Algorithm	PDs_1DP	$(2,0)$ batch k-means	$(2,0)$ k-means	$(2,0)$ smoka
# of iterations	na	7	1389	28
Quality	499,485	498,432	497,467	497,466
Algorithm	PDs_2DP	$(2,0)$ batch k-means	$(2,0)$ k-means	$(2,0)$ smoka
# of iterations	na	11	1519	36
Quality	499,044	498,092	497,334	497,385

dimensional approximations provided by PDs_2DP generate better partitions than those provided by PDs_1DP. The results also indicate that smoka generates clustering results of quality comparable with those generated by k-means.

The number of iterations performed by smoka is a *fraction* of the number of iterations performed by k-means. Note that distance computations are the most numerically expensive operation performed by k-means. In addition to computing the distances between high-dimensional vectors and centroids performed by k-means, each iteration of smoka also computes exponents of the distances. This additional computational effort is negligible as compared to that required to compute the distances. Moreover, for distance-like functions other than the quadratic Euclidean distance the incremental step may require as much computation as the batch step. In contrast smoka does not perform incremental iterations. In this case the computational cost of a single smoka iteration requires about 50% of the computational effort required by one iteration of the k-means algorithm.

In all experiments reported in this chapter smoka converges roughly as fast as batch k-means, and generates superior results. Convergence for $(2,0)$ k-means, $(2,0)$ batch k-means, and $(2,0)$ smoka applied to the initial partition generated by PDs_1DP is shown in Figure 4.5. While smoka stops after 28 iterations, already af-

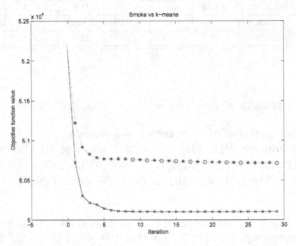

Fig. 4.5. Collection: Enron email dataset; quality Q_B of partitions generated by the first 28 iterations of $(2,0)$ k-means (batch iterations are marked by "*", incremental iterations are marked by "o"), and $(2,0)$ smoka (marked by "x").

ter the first eight iterations the smoka corresponding lower branch of the graph is almost flat.

Table 4.4 reports results for the purely logarithmic distance-like function

$$d(\mathbf{x}, \mathbf{y}) = \sum_{j=1}^{n} \left[\mathbf{y}[j] \log \frac{\mathbf{y}[j]}{\mathbf{x}[j]} + \mathbf{x}[j] - \mathbf{y}[j] \right].$$

In contrast to the quadratic case (reported in Table 4.3) PDs_1DP outperforms PDs_2DP. Nonetheless, the final partition generated by BIRCH, PDs_2DP, and k-means/smoka is of about the same quality as that generated by BIRCH, PDs_1DP, and k-means/smoka.

Finally Table 4.5 displays results for the logarithmic function augmented with the quadratic weight:

$$d(\mathbf{x}, \mathbf{y}) = \frac{50}{2} \|\mathbf{x} - \mathbf{y}\|^2 + \sum_{j=1}^{n} \left[\mathbf{y}[j] \log \frac{\mathbf{y}[j]}{\mathbf{x}[j]} + \mathbf{x}[j] - \mathbf{y}[j] \right].$$

Table 4.4. Collection: Enron email dataset; number of iterations and quality Q of $517,417$ vector dataset $\mathcal{A}$ partitions generated by (0,1) BIRCH, PDs_1DP/PDs_2DP, (0,1) k-means, and (0,1) smoka; the size of the dataset $\mathcal{B}$ is 6185, the vector space dimension is 5000, total of 10 clusters

Algorithm	PDs_1DP	(0, 1) batch k-means	(0, 1) k-means	(0, 1) smoka
# of iterations	na	34	2661	29
Quality	1.3723e+07	1.36203e+07	1.34821e+07	1.35365e+07
Algorithm	PDs_2DP	(0, 1) batch k-means	(0, 1) k-means	(0, 1) smoka
# of iterations	na	41	2717	33
Quality	1.37277e+07	1.36271e+07	1.34821e+07	1.35316e+07

Table 4.5. Collection: Enron email dataset; number of iterations and quality Q of $517,417$ vector dataset $\mathcal{A}$ 10 cluster partition partitions generated by (50,1) BIRCH, PDs_1DP/PDs_2DP, (50,1) k-means, and (50,1) smoka; the size of the dataset $\mathcal{B}$ is 5322, the vector space dimension is 5000, total of 10 clusters

Algorithm	PDs_1DP	(50, 1) batch k-means	(50, 1) k-means	(50, 1) smoka
# of iterations	na	2	4388	26
Quality	2.65837e+07	2.65796e+07	2.6329e+07	2.63877e+07
Algorithm	PDs_2DP	(50, 1) batch k-means	(50, 1) k-means	(50, 1) smoka
# of iterations	na	2	4371	27
Quality	2.6599e+07	2.659449e+07	2.63247e+07	2.63903e+07

In this case initial partitions generated through PDs_1DP are superior to those generated through PDs_2DP.

4.5 Conclusion

The chapter presented a three-step clustering procedure based on BIRCH, PDsDP, and k-means or smoka algorithms running with divergences. Our numerical experiments show that the fraction $\dfrac{\lambda_{j1} + \lambda_{j2}}{\Lambda_j}$ varies from 0.95 to 0.1 (here Λ_j is the total sum of the eigenvalues of cluster π_j, so that $\Lambda_j = Q(\pi_j)$ when $d(\mathbf{x}, \mathbf{y}) = \|\mathbf{x} - \mathbf{y}\|^2$). This observation suggests that N, the dimension of cluster π_j approximation generated by $PDs_ND P$, should be cluster dependent. This has the potential to improve the quality of final partition and, at the same time, to bring additional memory savings.

The trade-off between the computational cost of high-dimensional approximations and the quality of the final partitions will be investigated.

Good-quality BIRCH-generated clusters are paramount for consequent $PDs_N DP$ and k-means type clustering. A theoretical foundation for applications of BIRCH with Csiszár divergences is already available. The k-means algorithm with Csiszár divergence tends to build clusters with many words in common (see [KTN03], [TBD+06]). Application of BIRCH equipped with Csiszár divergence may lead to clusters with sparse centroids, and this approach will be examined.

In an attempt to speed up BIRCH clustering, one may consider selecting only a fraction of the existing clusters for an assignment of a "new" vector. A number of selection techniques have been reported recently in [Wia07].

The numerical experiments conducted by these authors indicate that smoka generates clustering results of quality comparable with those generated by k-means. At the same time the number of iterations performed by smoka is a *fraction* of the number of iterations performed by k-means. Moreover, the experiments show a sharp decrease in $\dfrac{\#\ of\ \texttt{smoka}\ iterations}{\#\ of\ k\text{-means iterations}}$ with the growing size of the dataset. Table 4.6 shows the ratio for some of the results reported in this section as well as for results reported in [TK05],[Kog07].

Table 4.6. Examples of iteration ratio per dataset size and distance-like function

Dataset size	Number of clusters	$\dfrac{(2,0)\ \texttt{smoka}}{(2,0)\ k\text{-means}}$ iteration ratio	$\dfrac{(0,1)\ \texttt{smoka}}{(0,1)\ k\text{-means}}$ iteration ratio	$\dfrac{(20,1)\ \texttt{smoka}}{(20,1)\ k\text{-means}}$ iteration ratio	$\dfrac{(50,1)\ \texttt{smoka}}{(50,1)\ k\text{-means}}$ iteration ratio
3891	3	0.08	0.04	0.02	
5322	10				0.0059
6101	10	0.033			
6185	10		0.01		
9469	20	0.015			
19997	20	0.0087	0.0029	0.0025	

Acknowledgment

Kogan's work was supported in part by the United States–Israel Binational Science Foundation (BSF).

References

[BB99]   M.W. Berry and M. Browne. *Understanding Search Engines: Mathematical Modeling and Text Retrieval*. SIAM, Philadelphia, 1999.

[Ber06]   P. Berkhin. A survey of clustering data mining techniques. In J. Kogan, C. Nicholas, and M. Teboulle, editors, *Grouping Multidimensional Data: Recent Advances in Clustering*, pages 25–72. Springer, Berlin, 2006.

[BFR98]   P.S. Bradley, U.M. Fayyad, and C. Reina. Scaling clustering algorithms to large databases. In *Proceedings of the Fourth International Conference on Knowledge Discovery and Data Mining*, pages 9–15, AAAI Press, Menlo Park, CA, 1998.

[BMDG04]   A. Banerjee, S. Merugu, I. S. Dhillon, and J. Ghosh. Clustering with Bregman divergences. In *Proceedings of the 2004 SIAM International Conference on Data Mining*, pages 234–245. SIAM, Philadelphia, 2004.

[Bol98]   D. Boley. Principal direction divisive partitioning. *Data Mining and Knowledge Discovery*, 2(4):325–344, 1998.

[CK99]   E. Chisholm and T. Kolda. New term weighting formulas for the vector space method in information retrieval, 1999. Report ORNL/TM-13756, Computer Science and Mathematics Division, Oak Ridge National Laboratory.

[DHS01]   R. Duda, P. Hart, and D. Stork. *Pattern Classification*. John Wiley & Sons, Inc., New York, second edition, 2001.

[DKN03]   I. S. Dhillon, J. Kogan, and C. Nicholas. Feature selection and document clustering. In M.W. Berry, editor, *Survey of Text Mining*, pages 73–100. Springer, New York, 2003.

[HLP34]   G. Hardy, J.E. Littlewood, and G. Polya. *Inequalities*. Cambridge University Press, Cambridge, 1934.

[HN01]   P. Hansen and N. Mladenovic. J-Means: a new local search heuristic for minimum sum of squares clustering. *Pattern Recognition*, 34:405–413, 2001.

[Kar99]   N.B. Karayiannis. An axiomatic approach to soft learning vector quantization and clustering. *IEEE Transactions on Neural Networks*, 10(5):1153–1165, 1999.

[KGV83]   S. Kirkpatrick, C.D. Gelatt, and M.P. Vecchi. Optimization by simulated annealing. *Science*, 220:671–680, 1983.

[KNV03]   J. Kogan, C. Nicholas, and V. Volkovich. Text mining with hybrid clustering schemes. In M.W.Berry and W.M. Pottenger, editors, *Proceedings of the Workshop on Text Mining (held in conjunction with the Third SIAM International Conference on Data Mining)*, pages 5–16, 2003.

[KNW07]   J. Kogan, C. Nicholas, and M. Wiacek. Hybrid clustering of large high dimensional data. In M. Castellanos and M.W. Berry, editors, *In Proceedings of the Workshop on Text Mining (held in conjunction with the SIAM International Conference on Data Mining)*. SIAM, 2007.

[Kog01]   J. Kogan. Means clustering for text data. In M.W.Berry, editor, *Proceedings of the Workshop on Text Mining at the First SIAM International Conference on Data Mining*, pages 47–54, 2001.

[Kog07]   J. Kogan. *Introduction to Clustering Large and High-Dimensional Data*. Cambridge University Press, New York, 2007.

[KT06]   J. Kogan and M. Teboulle. Scaling clustering algorithms with Bregman distances. In M.W. Berry and M. Castellanos, editors, *Proceedings of the Workshop on Text Mining at the Sixth SIAM International Conference on Data Mining*, 2006.

[KTN03]   J. Kogan, M. Teboulle, and C. Nicholas. The entropic geometric means algorithm: an approach for building small clusters for large text datasets. In D. Boley et al, editor, *Proceedings of the Workshop on Clustering Large Data Sets (held in conjunction with the Third IEEE International Conference on Data Mining)*, pages 63–71, 2003.

[LB06]   D. Littau and D. Boley. Clustering very large datasets with PDDP. In J. Kogan, C. Nicholas, and M. Teboulle, editors, *Grouping Multidimensional Data: Recent Advances in Clustering*, pages 99–126. Springer, New York, 2006.

[MRR⁺53] N. Metropolis, A.W. Rosenbluth, M.N. Rosenbluth, A.H. Teller, and E. Teller. Equations of state calculations by fast computing machines. *Journal of. Chemical Physics*, 21(6):1087–1091, 1953.

[NK95]   O. Nasraoui and R. Krishnapuram. Crisp interpretations of fuzzy and possibilistic clustering algorithms. In *Proceedings of 3rd European Congress on Intelligent Techniques and Soft Computing*, pages 1312–1318, ELITE Foundation, Aachen, Germany, April 1995.

[RGF90]   K. Rose, E. Gurewitz, and G. Fox. A deterministic annealing approach to clustering. *Pattern Recognition Letters*, 11(11):589–594, 1990.

[Ros98]   K. Rose. Deterministic annealing for clustering, compression, classification, regression, and related optimization problems. *Proceedings of the IEEE*, 86(11):2210–2239, 1998.

[Ste56]   H. Steinhaus. Sur la division des corps matèriels en parties. *Bulletin De L'Acadēmie Polonaise Des Sciences Classe III Mathematique, Astronomie, Physique, Chimie, Geologie, et Geographie*, 4(12):801–804, 1956.

[TBD⁺06] M. Teboulle, P. Berkhin, I. Dhillon, Y. Guan, and J. Kogan. Clustering with entropy-like k-means algorithms. In J. Kogan, C. Nicholas, and M. Teboulle, editors, *Grouping Multidimensional Data: Recent Advances in Clustering*, pages 127–160. Springer, Berlin, 2006.

[Teb07]   M. Teboulle. A unified continuous optimization framework for center-based clustering methods. *Journal of Machine Learning Research*, 8:65–102, 2007.

[TK05]   M. Teboulle and J. Kogan. Deterministic annealing and a k-means type smoothing optimization algorithm for data clustering. In I. Dhillon, J. Ghosh, and J. Kogan, editors, *Proceedings of the Workshop on Clustering High Dimensional Data and its Applications (held in conjunction with the Fifth SIAM International Conference on Data Mining)*, pages 13–22, SIAM, Philadelphia 2005.

[Wia07]   M. Wiacek. An implementation and evaluation of the balanced iterative reducing and clustering algorithm (BIRCH). Technical Report TR-CS-07-02, CSEE Department, UMBC, Baltimore, MD, March 2007.

[WS03]   S. Wang and D. Schuurmans. Learning continuous latent variable models with Bregman divergences. In *Lecture Notes in Artificial Intelligence*, volume 2842, pages 190–204, 2003. Available from World Wide Web: http: //www.cs.ualberta.ca/~dale/papers.html, http://www.cs. ualberta.ca/%7Eswang/publication/publication.html.

[Yu05]   J. Yu. General C-Means Clustering Model. *IEEE Transactions on Pattern Analysis and Machine Intelligence*, 27(8):1197–1211, 2005.

[ZHD99]   B. Zhang, M. Hsu, and U. Dayal. K-harmonic means—a data clustering algorithm. Technical Report HPL-1999-124 991029, HP Labs, Palo Alto, CA, 1999.

[ZKH99]   B. Zhang, G. Kleyner, and M. Hsu. A local search approach to K-clustering. HP Labs Technical Report HPL-1999-119, HP Labs, Palo Alto, CA, 1999. Available from World Wide Web: citeseer.ist.psu.edu/article/ zhang99local.html.

[ZRL97]   T. Zhang, R. Ramakrishnan, and M. Livny. BIRCH: A new data clustering algorithm and its applications. *Journal of Data Mining and Knowledge Discovery*,
1(2):141–182, 1997.

5

Text Clustering with Local Semantic Kernels

Loulwah AlSumait and Carlotta Domeniconi

Overview

Document clustering is a fundamental task of text mining, by which efficient organization, navigation, summarization, and retrieval of documents can be achieved. The clustering of documents presents difficult challenges due to the sparsity and the high dimensionality of text data, and to the complex semantics of natural language. Subspace clustering is an extension of traditional clustering that is designed to capture local feature relevance, and to group documents with respect to the features (or words) that matter the most.

This chapter presents a subspace clustering technique based on a locally adaptive clustering (LAC) algorithm. To improve the subspace clustering of documents and the identification of keywords achieved by LAC, kernel methods and semantic distances are deployed. The basic idea is to define a local kernel for each cluster by which semantic distances between pairs of words are computed to derive the clustering and local term weightings. The proposed approach, called *semantic LAC*, is evaluated using benchmark datasets. Our experiments show that semantic LAC is capable of improving the clustering quality.

5.1 Introduction

With the astonishing expansion of the internet, intranets, and digital libraries, billions of electronic text documents are made available. Extracting implicit, nontrivial, and useful knowledge from these huge corpora is essential for applications such as knowledge management in large business enterprises, semantic web, and automatic processing of messages, emails, surveys, and news.

Document clustering is a fundamental task of text mining by which efficient organization, navigation, summarization, and retrieval of documents can be achieved. Document clustering seeks to automatically partition unlabeled documents into groups. Ideally, such groups correspond to genuine themes, topics, or categories of the corpus [TSK06].

The classic document representation is a word-based vector, known as the vector space model (VSM) [STC04]. According to the VSM, each dimension is associated with one term from the dictionary of all the words that appear in the corpus. The VSM, although simple and commonly used, suffers from a number of deficiencies. Inherent shortages of the VSM include breaking multi-word expressions, like *machine learning*, into independent features, mapping synonymous words into different components, and treating polysemous as one single component. Although traditional preprocessing of documents, such as eliminating stop words, pruning rare words, stemming, and normalization, can improve the representation, it is still essential to embed semantic information and conceptual patterns in order to enhance the prediction capabilities of clustering algorithms.

Moreover, the VSM representation of text data can easily result in hundreds of thousands of features. As a consequence, any clustering algorithm would suffer from the *curse of dimensionality*. In such sparse and high-dimensional space, any distance measure that assumes all features to have equal importance is likely to be ineffective. This is because points within the same cluster would at least have a few dimensions on which they are far apart from each other. As a result, the farthest point is expected to be almost as close as the nearest one. Since feature relevance is local in nature (e.g., a single word may have a different importance across different categories), global feature selection approaches are not effective, and may cause a loss of crucial information. To capture local feature relevance, local operations that embed different distance measures in different regions are required. Subspace clustering is an extension of traditional clustering that is designed to group data points, that is, documents, with respect to the features (or words) that matter the most.

In this chapter, a subspace clustering technique based on the locally adaptive clustering (LAC) algorithm [DGM$^+$07] is used. To improve the subspace clustering of documents and the identification of keywords achieved by LAC, kernel methods and semantic distances are deployed. The idea is to define a local kernel for each cluster, by which semantic distances between pairs of words are computed to derive the clustering and local term weightings.

The chapter is organized as follows. Some background on kernel methods for text is provided in Section 5.2. The LAC algorithm is briefly described in Section 5.3. In Sections 5.4 and 5.5, we present our approach, the experimental design, and results. A review of recent work on the application of semantic information in kernel-based learning methods is provided in Section 5.6. Our final conclusions and suggestions for future work are discussed in Section 5.7.

5.2 Kernel Methods for Text

Kernel methods are a promising approach to pattern analysis. Formally, the VSM can be defined as the following mapping:

$$\phi : d \mapsto \phi(d) = (tf(t_1, d), tf(t_2, d), \dots, tf(t_D, d)) \in \mathcal{R}^D$$

where $tf(t_i, d)$ is the frequency of term t_i in document d, and D is the size of the dictionary.

To represent the whole corpus of N documents, the document-term matrix , $\mathcal{D}$, is introduced. $\mathcal{D}$ is a $N \times D$ matrix whose rows are indexed by the documents and whose columns are indexed by the terms [STC04].

The basic idea of kernel methods is to embed the data in a suitable feature space, such that solving the problem in the new space is easier (e.g., linear). A kernel represents the similarity between two objects (e.g., documents or terms), defined as dot-product in this new vector space. The kernel trick allows keeping the mapping implicit. In other words, it is only required to know the inner products between the images of the data items in the original space. Therefore, defining a suitable kernel means finding a good representation of the data objects.

In text mining, semantically similar documents should be mapped to nearby positions in feature space. To address the omission of semantic content of the words in the VSM, a transformation of the document vector of the type $\tilde{\phi}(d) = \phi(d)Sem$ is required, where Sem is a semantic matrix. Different choices of the matrix Sem lead to different variants of the VSM. Using this transformation, the corresponding vector space kernel takes the form

$$\tilde{k}(d_1, d_2) = \phi(d_1)SemSem^\top \phi(d_2)^\top \qquad (5.1)$$
$$= \tilde{\phi}(d_1)\tilde{\phi}(d_2)^\top$$

Thus, the inner product between two documents d_1 and d_2 in feature space can be computed efficiently directly from the original data items using a kernel function. The semantic matrix Sem can be created as a composition of successive embeddings, which add additional refinements to the semantics of the representation. Therefore, Sem can be defined as:

$$Sem = RP$$

where R is a diagonal matrix containing the term weightings or relevances, while P is a *proximity matrix* defining the semantic similarities between the different terms of the corpus. One simple way of defining the term-weighting matrix R is to use the inverse document frequency (idf) . In this chapter, a new weighting measure, dynamically learned by means of the LAC algorithm, is used to construct R.

P has nonzero off-diagonal entries, $P_{ij} > 0$, when the term i is semantically related to the term j. Embedding P in the vector space kernel corresponds to representing a document as a less sparse vector, $\phi(d)P$, which has nonzero entries for all terms that are semantically similar to those present in document d. There are different methods for obtaining P. A semantic network, like *WordNet*, which encodes relationships between words of a dictionary in a hierarchical fashion, is one source of term similarity information. An alternative, the proximity matrix, can be computed using *latent semantic indexing (LSI)*. The singular value decomposition (SVD) of the matrix $\mathcal{D}^\top$ is calculated to extract the semantic information, and to project the documents onto the space spanned by the first k eigenvectors of the $\mathcal{D}^\top \mathcal{D}$ matrix. The corresponding kernel is called latent semantic kernel (LSK). The simplest method to compute P, and used in this chapter, is the generalized vector space model (GVSM) [WZW85]. This technique aims at capturing correlations of terms by investigating their co-occurrences across the corpus. Two terms are considered semantically related if they frequently co-occur in the same documents. Thus, a document is represented by the embedding

$$\tilde{\phi}(d) = \phi(d)\mathcal{D}^\top$$

and the corresponding kernel is

$$\tilde{k}(d_1, d_2) = \phi(d_1)\mathcal{D}^\top \mathcal{D}\phi(d_2)^\top \tag{5.2}$$

where the (i, j)th entry of the matrix $\mathcal{D}^\top \mathcal{D}$ is given by

$$(\mathcal{D}^\top \mathcal{D})_{ij} = \sum_d tf(t_i, d)tf(t_j, d)$$

The matrix $\mathcal{D}^T \mathcal{D}$ has a nonzero entry $(\mathcal{D}^T \mathcal{D})_{ij}$ if there is a document d in which the corresponding terms t_i and t_j co-occur, and the strength of the relationship is given by the frequency and the number of their co-occurrences.

5.3 Locally Adaptive Clustering (LAC)

As mentioned earlier, clustering suffers from the curse of dimensionality problem in high-dimensional spaces. Furthermore, several clusters may exist in different subspaces, comprised of different combinations of features. In many real-world problems, in fact, some points are correlated with respect to a given set of dimensions, and others are correlated with respect to different dimensions. Each dimension could be relevant to at least one of the clusters.

To capture the local correlations of data, a proper feature selection procedure should operate locally in the input space. Local feature selection allows one to embed different distance measures in different regions of the input space; such distance metrics reflect local correlations of data. LAC [DGM+07] (a preliminary version appeared in [DPGM04]) is a soft feature selection procedure that assigns (local) weights to features. Dimensions along which data are loosely correlated receive a small weight, which has the effect of elongating distances along that dimension. Features along which data are strongly correlated receive a large weight, which has the effect of constricting distances along that dimension. Figure 5.1 gives a simple example. The upper plot depicts two clusters of data elongated along the x and y dimensions. The lower plot shows the same clusters, where within-cluster distances between points are computed using the respective local weights generated by our algorithm. The weight values reflect local correlations of data, and reshape each cluster as a dense spherical cloud. This directional local reshaping of distances better separates clusters, and allows for the discovery of different patterns in different subspaces of the original input space.

Thus, LAC discovers *weighted clusters*. A weighted cluster is a subset of data points, together with a weight vector **w**, such that the points are closely clustered

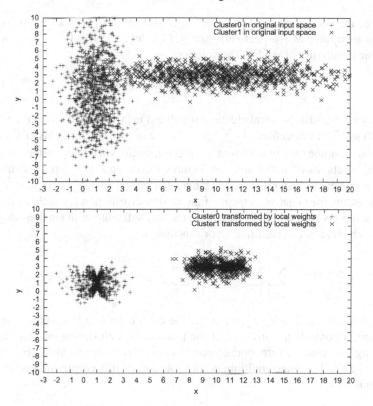

Fig. 5.1. (Top) Clusters in original input space. (Bottom) Clusters transformed by local weights.

according to the corresponding weighted Euclidean distance. The w_j measures the degree of participation of feature j to the cluster. The objective of LAC is to find cluster centroids and weight vectors.

The partition induced by discovering weighted clusters is formally defined as follows. Given a set S of N points $\mathbf{x} \in \mathcal{R}^D$, a set of k centers $\{\mathbf{c}_1, ..., \mathbf{c}_k\}, \mathbf{c}_j \in \mathcal{R}^D$, $j = 1, ..., k$, coupled with a set of corresponding weight vectors $\{\mathbf{w}_1, ..., \mathbf{w}_k\}, \mathbf{w}_j \in \mathcal{R}^D, j = 1, ..., k$, partition S into k sets:

$$S_j = \{\mathbf{x} | (\sum_{i=1}^{D} w_{ji}(x_i - c_{ji})^2)^{\frac{1}{2}} < (\sum_{i=1}^{D} w_{qi}(x_i - c_{qi})^2)^{\frac{1}{2}}, \forall q \neq j\}$$

where w_{ji} and c_{ji} represent the ith components of vectors $\mathbf{w}_j$ and $\mathbf{c}_j$, respectively.

The set of centers and weights is optimal with respect to the Euclidean norm, if they minimize the error measure:

$$E_1(C, W) = \sum_{j=1}^{k} \sum_{i=1}^{D} (w_{ji} \frac{1}{|S_j|} \sum_{\mathbf{x} \in S_j} (c_{ji} - x_i)^2)$$

subject to the constraints $\forall j \sum_i w_{ji} = 1$. C and W are $(D \times k)$ matrices whose columns are $\mathbf{c}_j$ and $\mathbf{w}_j$, respectively, that is, $C = [\mathbf{c}_1...\mathbf{c}_k]$ and $W = [\mathbf{w}_1...\mathbf{w}_k]$, and $|S_j|$ is the cardinality of set S_j. The solution

$$(C^*, W^*) = \arg\min_{C,W} E_1(C, W)$$

will discover one-dimensional clusters: it will put maximal (unit) weight on the feature with smallest dispersion $\frac{1}{|S_j|} \sum_{\mathbf{x} \in S_j} (c_{ji} - x_i)^2$, within each cluster j, and zero weight on all other features. To find weighted multidimensional clusters, where the unit weight gets distributed among all features according to the respective dispersion of data within each cluster, LAC adds the regularization term $\sum_{i=1}^{D} w_{ji} \log w_{ji}$. This term represents the negative entropy of the weight distribution for each cluster. It penalizes solutions with maximal weight on the single feature with smallest dispersion within each cluster. The resulting error function is

$$E(C, W) = \sum_{j=1}^{k} \sum_{i=1}^{D} (w_{ji} \frac{1}{|S_j|} \sum_{\mathbf{x} \in S_j} (c_{ji} - x_i)^2 + h w_{ji} \log w_{ji})$$

subject to the constraints $\forall j \sum_i w_{ji} = 1$. The coefficient $h \geq 0$ is a parameter of the procedure; it controls the strength of the incentive for clustering on more features. Increasing its value will encourage clusters on more features, and vice versa. By introducing the Lagrange multipliers, the solution of this constrained optimization problem is

$$w_{ji}^* = \frac{\exp\left(-\frac{1}{|S_j|} \sum_{\mathbf{x} \in S_j} (c_{ji} - x_i)^2 \Big/ h\right)}{\sum_{i=1}^{D} \exp\left(-\frac{1}{|S_j|} \sum_{\mathbf{x} \in S_j} (c_{ji} - x_i)^2 \Big/ h\right)} \quad (5.3)$$

$$c_{ji}^* = \frac{1}{|S_j|} \sum_{\mathbf{x} \in S_j} x_i \quad (5.4)$$

To find a partition that identifies the clustering solution, a search strategy that progressively improves the quality of initial centroids and weights is proposed. The search starts with *well-scattered* points in S as the k centroids, and weights equally set to $1/D$. Then, for each centroid $\mathbf{c}_j$, the corresponding sets S_j are computed as previously defined. Next, the average distance of the points in S_j to the centroid $\mathbf{c}_j$, along each dimension, is computed. The smaller the average, the stronger the degree of participation of feature i to cluster j. To credit weights to features (and to clusters), the average distances are used, as given above. Consequently, the computed weights are used to update the sets S_j, and therefore the centroids' coordinates. The procedure is iterated until convergence is reached.

5.4 Semantic LAC

The local weights provided by LAC are exploited to identify the keywords specific to each topic (cluster). To improve the subspace clustering of documents and the computation of local term weights, the learning paradigm of kernel methods is used. The idea is to use a semantic distance between pairs of words by defining a local kernel for each cluster, derived from Equation (5.1), as follows:

$$K_j(d_1, d_2) = \phi(d_1) Sem_j Sem_j^\top \phi(d_2)^\top, \qquad (5.5)$$

where $Sem_j = R_j P$. R_j is a local term-weighting diagonal matrix corresponding to cluster j, and P is the proximity matrix between the terms. The weight vector that LAC learns is used to construct the weight matrix R for each cluster. Formally, R_j is a diagonal matrix where $r_{ii} = w_{ji}$, that is, the weight of term i for cluster j, for $i = 1, \ldots, D$, and can be expressed as follows:

$$R_j = \begin{pmatrix} w_{j1} & 0 & \ldots & 0 \\ 0 & w_{j2} & \ldots & 0 \\ \vdots & \vdots & \vdots & \vdots \\ 0 & 0 & \ldots & w_{jD} \end{pmatrix}$$

To compute P, the GVSM is used. Since P holds a similarity figure between words in the form of co-occurrence information, it is necessary to transform it to a distance measure before utilizing it. To this end, the values of P are recomputed as follows:

$$P_{ij}^{dis} = 1 - (P_{ij}/max(P)),$$

where $max(P)$ is the maximum entry value in the proximity matrix.

Consequently, a *semantic dissimilarity matrix* for cluster j is a $D \times D$ matrix given by

$$\widehat{Sem}_j = R_j P^{dis},$$

which represents semantic dissimilarities between the terms with respect to the local term weightings. Thus, by means of local kernels driven by dissimilarity matrices, the new representation of documents highlights terms or dimensions that have higher degree of relevance for the corresponding cluster, as well as terms that are semantically similar.

Such transformation enables the kernels to be plugged-in directly in the dynamic of LAC. This results in our proposed *semantic LAC* algorithm.

5.4.1 Semantic LAC Algorithm

Similarly to the LAC algorithm described earlier in Section 5.3, semantic LAC starts with k initial centroids and equal weights. It partitions the data points, recomputes the weights and data partitions accordingly, and then recomputes the new centroids. The algorithm iterates until convergence, or a maximum number of iterations is exceeded.

Semantic LAC uses a semantic distance. A point $\mathbf{x}$ is assigned to the cluster j that minimizes the semantic distance of the point from its centroid. The semantic distance is derived from the kernel in Eq. (5.5) as follows:

$$L_w(\mathbf{c}_j, \mathbf{x}) = (\mathbf{x} - \mathbf{c}_j)\widehat{Sem}_j\widehat{Sem}_j^{\top}(\mathbf{x} - \mathbf{c}_j)^{\top}.$$

Thus, every time the algorithm computes S_j's, the semantic matrix must be computed by means of the new weights. The resulting algorithm, called semantic LAC, is summarized in Algorithm 5.4.1.

Algorithm 5.4.1 Semantic LAC

Input: N points $\mathbf{x} \in \mathcal{R}^D$, k, and h
1. Initialize k centroids $\mathbf{c}_1, \mathbf{c}_2, \ldots, \mathbf{c}_k$
2. Initialize weights: $w_{ji} = \frac{1}{D}$, for each centroid $\mathbf{c}_j, j = 1, \ldots, k$,
   and for each feature $i = 1, \ldots, D$
3. Compute P; then compute P^{dis}
4. Compute $\widehat{Sem}$ for each cluster j: $\widehat{Sem}_j = R_j P^{dis}$
5. For each centroid $\mathbf{c}_j$, and for each point $\mathbf{x}$, set:
   $S_j = \{\mathbf{x} | j = \arg\min_l L_w(\mathbf{c}_l, \mathbf{x})\}$,
   where $L_w(\mathbf{c}_l, \mathbf{x}) = (\mathbf{x} - \mathbf{c}_l)\widehat{Sem}_l\widehat{Sem}_l^{\top}(\mathbf{x} - \mathbf{c}_l)^{\top}$
6. Compute new weights:
   for each centroid $\mathbf{c}_j$, and for each feature i:
$$w_{ji}^* = \frac{\exp\left(-\frac{1}{|S_j|}\sum_{\mathbf{x} \in S_j}(c_{ji} - x_i)^2 \Big/ h\right)}{\sum_{i=1}^{D}\exp\left(-\frac{1}{|S_j|}\sum_{\mathbf{x} \in S_j}(c_{ji} - x_i)^2 \Big/ h\right)};$$
7. For each centroid $\mathbf{c}_j$:
   Recompute $\widehat{Sem}_j$ matrix using new weights;
8. For each point $\mathbf{x}$:
   Recompute $S_j = \{\mathbf{x} | j = \arg\min_l L_w(\mathbf{c}_l, \mathbf{x})\}$
9. Compute new centroids:
   $\mathbf{c}_j = \frac{\sum_{\mathbf{x}} \mathbf{x} 1_{S_j}(\mathbf{x})}{\sum_{\mathbf{x}} 1_{S_j}(\mathbf{x})}$ for each $j = 1, \ldots, k$,
   where $1_s(.)$ is the indicator function of set S
10. Iterate 5–9 until convergence, or
    maximum number of iterations is exceeded

The running time of one iteration of LAC is $O(kND)$, where k is the number of clusters, N is the number of data, and D is the number of dimensions. For semantic LAC, local kernels are computed for each cluster, based on semantic distances between pairs of terms. Thus, the running time of one iteration becomes $O(kND^2)$. However, we perform feature selection in our experiments, and reduce D to the hundreds. In addition, as discussed later, semantic LAC reaches better solutions in fewer iterations than LAC in general. Thus, by stopping the execution of semantic LAC when a maximum number of iterations is reached, we are able to limit the computational burden without affecting accuracy.

5.5 Experimental Results

5.5.1 Datasets

The datasets used in our experiments (see Table 5.1) were preprocessed according to the following steps: removal of stop words, stemming of words to their root source, and removal of rare words that appeared in less than four documents. A global feature selection algorithm, called DocMine, which is based on frequent itemset mining, was also performed [BDK04]. Briefly, DocMine mines the documents to find the frequent itemsets, which are sets of words that co-occur frequently in the corpus, according to a given support level (SL). In principle, the support level is driven by the target dimensionality of the data. The union of such frequent items is used to represent each document as a bag of frequent itemsets. The weight of the new entry is the frequency of the corresponding word in the document [KDB05, BDK04]. In the following, we provide a short description of the datasets.

Email-1431. The original Email-1431 corpus consists of texts from 1431 emails manually classified into three categories: conference (370), job (272), and spam (789). The total dictionary size is 38713 words. In this chapter, we consider a two-class classification problem by combining the conference and job emails into one class (NS = Non Spam). In addition, the set with 285 features, that corresponds to 10% support level, is used.

Ling-Spam. This dataset is a mixture of spam messages (453) and messages (561) sent via the linguist list, a moderated list concerning the profession and science of linguistics. The original size of the dictionary is 24627. In our experiments, the sets with 350, 287, 227, and 185 features were used, corresponding to 7%, 8%, 9%, and 10% support level, respectively.

20NewsGroup. This dataset is a collection of 20,000 messages collected from 20 different net-news newsgroups. In this chapter, two-class classification problems are considered using the following two categories: auto (990 documents) and space (987 documents); electronics (981 documents) and medical (990 documents). The dimension of the former set is 166, which correspond to 5% support level, while the latter set has 134 features with 5% support level.

Classic3. This dataset is a collection of abstracts from three categories: MEDLINE (1033 abstracts from medical journals), CISI (1460 abstracts from IR papers), and CRANFIELD (1399 abstracts from aerodynamics papers). We consider four problems constructed from the Classic3 set, which consist of 584 $(SL = 2\%)$, 395 $(SL = 3\%)$, 277 $(SL = 4\%)$, and 219 $(SL = 5\%)$ features, respectively.

5.5.2 Results

We ran LAC six times on all the datasets for $1/h = 1, \ldots, 6$. Table 5.2 lists average error rates, standard deviations, and minimum error rates obtained by running

Table 5.1. Characteristics of the datasets

Dataset	k	SL	D	N	points/class
Email1431	2	10	285	1431	S(789), NS(642)
Ling-Spam(10%)	2	10	185	1014	S(453), NS(561)
Ling-Spam(9%)	2	9	227	1014	S(453), NS(561)
Ling-Spam(8%)	2	8	287	1014	S(453), NS(561)
Ling-Spam(7%)	2	7	350	1014	S(453), NS(561)
Auto-Space	2	5	166	1977	A(990), S(987)
Medical-Elect.	2	5	134	1971	M(990), E(981)
Classic3(5%)	3	5	219	3892	Med(1033), Cran(1399), Cisi(1460)
Classic3(4%)	3	4	277	3892	Med(1033), Cran(1399), Cisi(1460)
Classic3(3%)	3	3	395	3892	Med(1033), Cran(1399), Cisi(1460)
Classic3(2%)	3	2	584	3892	Med(1033), Cran(1399), Cisi(1460)

Table 5.2. Experimental results of semantic LAC, LAC, and k-means

	LAC clustering			Sem LAC clustering			k-means clustering		
Dataset	Ave error	Std dev	Min error	Ave error	Std dev	Min error	Ave error	Std dev	Min error
Email1431	2.12	0.45	1.54	1.7	0.21	1.50	42.85	3.95	40.01
Ling-Spam 10%	6.33	0.28	5.96	4.67	0.26	4.24	20.07	19.16	6.31
Ling-Spam 9%	10.8	12.7	5.3	3.63	0.08	3.55	23.47	19.78	8.97
Ling-Spam 8%	5.5	0.9	4.0	3.18	0.18	3.06	20.15	18.74	7.40
Ling-Spam 7%	12.2	12.1	5.2	3.1	3.06	5.4	31.69	19.09	6.71
Auto-Space	28.7	6.35	24.7	27.6	2.02	25.0	42.85	3.95	40.01
Medical-Elect.	27.47	4.37	24.7	25.67	2.61	24.35	44.83	2.96	42.89
Classic3 5%	24.54	10.97	12.25	10.79	0.32	10.38	27.22	11.44	12.79
Classic3 4%	10.18	0.81	9.0	9.36	0.33	8.99	29.05	9.24	10.63
Classic3 3%	18.28	8.49	7.24	8.46	0.45	8.04	28.44	9.59	9.27
Classic3 2%	11.8	7.3	5.9	7.15	0.5	6.45	23.03	16.13	8.58

semantic LAC on all the datasets, along with the corresponding results of the LAC algorithm, and k-means as baseline comparison. Figures 5.2 and 5.3 illustrate the error rates of semantic LAC and LAC as a function of the h parameter values for the Classic3 (3%) and NewsGroup/Medical-Electronic, respectively. Error rates are computed according to the confusion matrices using the ground truth labels.

LAC and semantic LAC provided superior partitionings of the data with respect to k-means for all the datasets. As a further enhancement, semantic LAC provided error rates lower than LAC for all the datasets, many of which, as in Ling-Spam and Classic3, are with major improvements. Although, in some cases, LAC found solutions with lowest minimum error rates, for example, Newsgroup/Auto-Space and Classic3 2% and 3%, semantic LAC, on average, performed better. In addition, the standard deviations of the error rates for semantic LAC were significantly smaller than those of LAC, which demonstrates the stability of our subspace clustering

approach when semantic information is embedded. The robustness of semantic LAC
with respect to the parameter h is clearly depicted in Figures 5.2 and 5.3. This is a
relevant result since the setting of the h parameter is an open problem, as no domain
knowledge for its tuning is likely to be available in practice. Furthermore, parameter tuning is a difficult problem for clustering in general. Thus, the achievement of
robust clustering is a highly desirable result.

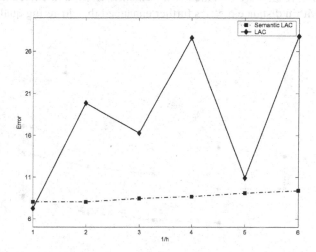

Fig. 5.2. Error rate vs. parameter h for semantic LAC and LAC on Classic3 3%.

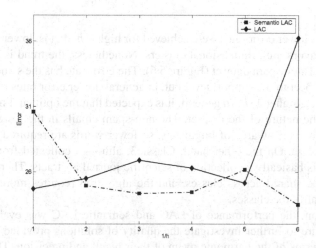

Fig. 5.3. Error rate vs. parameter h for semantic LAC and LAC on NewsGroup20/Medical-
Electronics.

We further investigated the behavior of semantic LAC using different support levels. Figures 5.4 and 5.5 illustrate the error rate vs. h values for different support levels on the Classic3 data, using semantic LAC and LAC, respectively. For increasing support values, that is, decreasing number of selected features, semantic LAC provided higher error rates. At the same time, no clear trend can be observed for LAC (see Figure 5.5). As expected, as more features are used to represent the documents, better clustering results are obtained with semantic LAC, and embedding semantic information within distance metrics further enhanced the clustering quality.

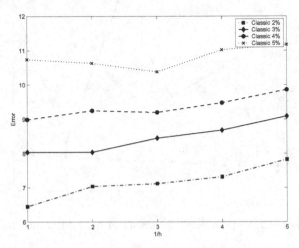

Fig. 5.4. Error rate vs. parameter h for different support levels on Classic3 using semantic LAC.

Moreover, lower error rates were achieved for higher h, that is, lower $1/h$ (Figure 5.4), which favors multidimensional clusters. Nonetheless, the trend is slightly different for the Ling-Spam dataset (Figure 5.6). The error rate has the same increasing trend with respect to the support level but, in general, lower error rates resulted from lower h, that is, higher $1/h$. In general, it is expected that the optimal dimensionality depends on the nature of the dataset. The non-spam emails in the Ling-Spam data come from one narrow area of linguistics, so fewer words are required to correctly identify the class. On the other hand, Classic3, although collected from three different areas, is basically a collection of scientific journal abstracts. Therefore, many words may be shared across classes, and the algorithm requires more features to correctly identify the classes.

In addition, the performance of LAC and semantic LAC was evaluated using the F1 measure to further investigate the quality of solutions provided by both approaches in terms of the harmonic mean of their recall and precision. The recall for a given class is the fraction of documents of the class that was correctly clustered in one group, over the total number of documents in the class. On the other hand, the precision for a given class is the fraction of documents of the class that was correctly

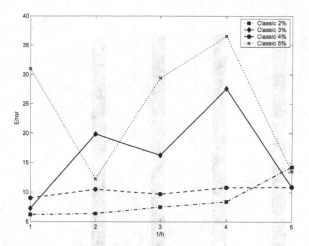

Fig. 5.5. Error rate vs. parameter h for different support levels on Classic3 using LAC.

clustered in one group, over the total number of documents assigned to that group. Figure 5.7 shows the F1 measure, averaged over the six runs corresponding to different h values, for semantic LAC and LAC on the four sets of the Ling-Spam data. The exact F1 measures for different h values for the Ling-Spam 9% data set is shown in Figure 5.8. It can be seen that the recall and precision of semantic LAC are higher than those of LAC.

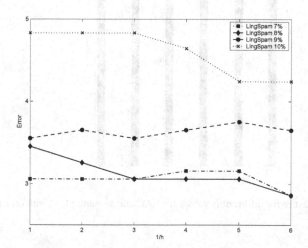

Fig. 5.6. Error rate vs. parameter h for different support levels on Ling-Spam data using semantic LAC.

Finally, we compared the convergence behavior of semantic LAC and LAC. Results are shown in Figures 5.9, 5.10, and 5.11. To partially eliminate the randomness

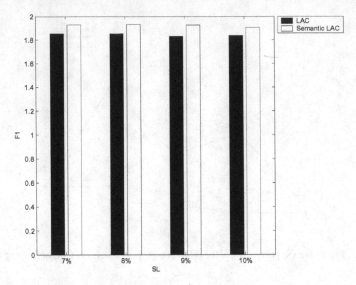

Fig. 5.7. Average F1 measure for LAC and semantic LAC run on Ling-Spam sets.

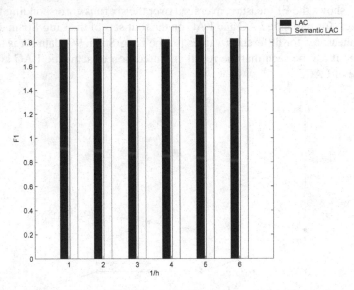

Fig. 5.8. F1 measure for different h values for LAC and semantic LAC run on Ling-Spam 9%.

of the initial centroids, the figures plot the error rate starting at iteration three for both LAC and semantic LAC. Figure 5.9 shows the error rate at every iteration of a single run for the Classic3 5% set. Figures 5.10 and 5.11 illustrate the error rate at each iteration averaged over six runs for the Ling-Spam 10% and the Medical-Electronic datasets, respectively. Although one iteration of semantic LAC requires more computations, semantic LAC does converge to a stable error value in fewer iterations. In fact, on average, semantic LAC reaches better error rates in less than six iterations. In particular, the results reported in Table 5.2 are executed with the maximum threshold for the number of iterations set to five. This demonstrates the potential of using local kernels that embed semantic relations as a similarity measure.

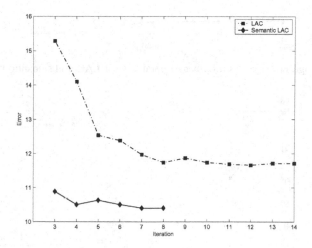

Fig. 5.9. Error rate vs. number of iterations for LAC and semantic LAC on Classic3 5%.

5.6 Related Work

There are a number of subspace clustering algorithms in the literature of data mining. Examples of such algorithms include the CLIQUE algorithm [AGGR98], density-based optimal projective clustering (DOC) [PJAM02], and projected clustering (PROCLUS) [AWY+99]. A comparative survey of subspace clustering techniques can be found in [PHL04].

The problem of embedding semantic information within the document representation and/or distance metrics has recently attracted a lot of attention. Most of this work focuses on text classification [BCM06, CSTL02, LY05]. In particular, Cristianini et al. [CSTL02] introduced and developed latent semantic kernels (LSK), as described earlier, and applied their novel approach on multiclass text classification problems. LSKs were tested using Reuters21578, and some improvement was

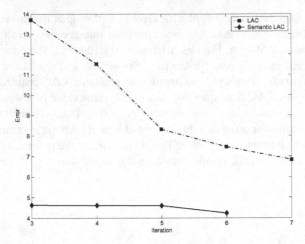

Fig. 5.10. Average error rate vs. number of iterations for LAC and semantic LAC on Ling-Spam 10%.

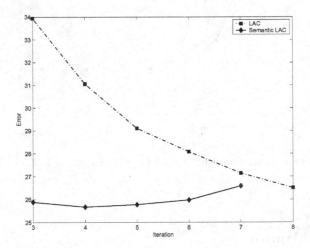

Fig. 5.11. Average error rate vs. number of iterations for LAC and semantic LAC on Medical-Electronics.

reported. The authors in [SdB00] have combined a semantic kernel with SVMs to define a new similarity measure for text classification. To identify semantic proximities between words, WordNet was utilized: the length of the path linking each possible pair of words in the taxonomy was used to measure the similarity. By incorporating the proximities within the VSM, documents were represented by new less sparse vectors and, hence, a new distance metric was induced and integrated into K-NN and SVMs. While the work in [SdB00] reported improvement in terms of accuracy when the semantic kernel was deployed, the task involves supervised learning.

Hotho et al. [HSS03] integrated conceptual account of terms found in WordNet to investigate its effects when deployed for unsupervised document clustering. They introduced different strategies for disambiguation, and applied bi-section k-means, a variant of k-means clustering [SKK00], with the cosine similarity measure on the Reuters21578 corpus. Similarly, [WH06] deployed WordNet to define a sense disambiguation method based on the semantic relatedness among the senses that was used in basic document clustering algorithms, for example, k-means, bisecting k-means, and HAC. They found that the use of senses of words can improve the clustering performance, but the improvement was statistically insignificant.

Recently [JZNH06], a subspace clustering approach that uses an ontology-based semantic distance has been proposed. In this approach, an ontology-based VSM is constructed from the original term-based VSM by means of a principal-component-analysis–like factorization of a term mutual information matrix. This approach captures the semantic relationship between each pair of terms based on the WordNet ontology. Similarly to LAC, the work in [JZNH06] applied the new representation to a subspace clustering algorithm, extended from the standard k-means. It identifies clusters by assigning large weights to the features that form the subspaces in which the clusters exist. The work in [JZNH06] generates fixed term weights prior to the clustering process. On the other hand, our semantic distance measure is driven by the local term weighting within each cluster, and the clusters together with the local weights are derived and enhanced by the embedded semantic, iteratively during the clustering process.

5.7 Conclusions and Future Work

In this chapter, the effect of embedding semantic information within subspace clustering of text documents was investigated. In particular, a semantic distance based on a GVSM kernel approach is embedded in a locally adaptive clustering algorithm to enhance the subspace clustering of documents, and the identification of relevant terms. Results have shown improvements over the original LAC algorithm in terms of error rates for all datasets tested. In addition, the semantic distances resulted in more robust and stable subspace clusterings.

The proposed approach can be explored further. In particular, in our future work we plan to perform more experiments using different datasets and various feature selection approaches. In addition, other kernel methods, for example, semantic smoothing of the VSM, LSK, and diffusion kernels, may provide more sophisticated semantic representations. Furthermore, an analysis of the distribution of the terms' weights produced by semantic LAC may identify the keywords that best represent the semantic topics discussed in the documents.

Acknowledgment

This work was supported in part by National Science Foundation (NSF) CAREER Award IIS-0447814.

References

[AGGR98] R. Agrawal, J. Gehrke, D. Gunopulos, and P. Raghavan. Automatic subspace clustering of high dimensional data for data mining applications. In *Proceedings of the ACM SIGMOD International Conference on Management of Data*, pages 94–105, ACM Press, New York, 1998.

[AWY⁺99] C.C. Aggarwal, J.L. Wolf, P.S. Yu, C. Procopiuc, and J.S. Park. Fast algorithms for projected clustering. In *Proceedings of the ACM SIGMOD International Conference on Management of Data*, pages 61–72, ACM Press, New York, 1999.

[BCM06] R. Basili, M. Cammisa, and A. Moschitti. A semantic kernel to classify texts with very few training examples. *Informatica*, 30:163–172, 2006.

[BDK04] D. Barbará, C. Domeniconi, and N. Kang. Classifying documents without labels. In *Proceedings of the Fourth SIAM International Conference on Data Mining*, pages 502–506, SIAM, Philadelphia, 2004.

[CSTL02] N. Cristianini, J. Shawe-Taylor, and H. Lodhi. Latent semantic kernels. *Journal of Intelligent Information Systems*, 18(2-3):127–152, 2002.

[DGM⁺07] C. Domeniconi, D. Gunopulos, S. Ma, B. Yan, M. Al-Razgan, and D. Papadopoulos. Locally adaptive metrics for clustering high dimensional data. *Data Mining and Knowledge Discovery Journal*, 14(1):63–97, 2007.

[DPGM04] C. Domeniconi, D. Papadopoulos, D. Gunopulos, and S. Ma. Subspace clustering of high dimensional data. In *Proceedings of the Fourth SIAM International Conference on Data Mining*, pages 517–521, SIAM, Philadelphia, 2004.

[HSS03] A. Hotho, S. Staab, and G. Stumme. Wordnet improves text document clustering. In *Proceedings of the Workshop on Semantic Web, SIGIR-2003*, Toronto, Canada, 2003.

[JZNH06] L. Jing, L. Zhou, M.K. Ng, and J. Zhexue Huang. Ontology-based distance measure for text clustering. In *Proceedings of the Text Mining Workshop, SIAM International Conference on Data Mining*, SIAM, Philadelphia, 2006.

[KDB05] N. Kang, C. Domeniconi, and D. Barbará. Categorization and keyword identification of unlabeled documents. In *Proceedings of the Fifth International Conference on Data Mining*, pages 677–680, IEEE, Los Alamitos, CA, 2005.

[LY05] C.-H. Lee and H.-C. Yang. A classifier-based text mining approach for evaluating semantic relatedness using support vector machines. In *Proceedings of the IEEE International Conference on Information Technology: Coding and Computing (ITCC'05)*, pages 128–133, IEEE, Los Alamitos, CA, 2005.

[PHL04] L. Parsons, E. Haque, and H. Liu. Evaluating subspace clustering algorithms. In *Proceedings of the Fourth SIAM International Conference on Data Mining*, pages 48–56, SIAM, Philadelphia, 2004.

[PJAM02] C.M. Procopiuc, M. Jones, P.K. Agarwal, and T.M. Murali. A monte carlo algorithm for fast projective clustering. In *Proceedings of the ACM SIGMOD International Conference on Management of Data*, pages 418–427, ACM Press, New York, 2002.

[SdB00] G. Siolas and F. d'Alché Buc. Support vector machines based on a semantic kernel for text categorization. In *Proceedings of the International Joint Conference on Neural Networks (IJCNN'00)*, pages 205–209, IEEE, Los Alamitos, CA, 2000.

[SKK00] M. Steinbach, G. Karypis, and V. Kumar. A comparison of document clustering techniques. In *Proceedings of the Sixth ACM SIGKDD World Text Mining Conference*, Boston, MA, 2000. Available from World Wide Web: `citeseer.nj.nec.com/steinbach00comparison.html`.

[STC04]   J. Shawe-Taylor and N. Cristianini. *Kernel Methods for Pattern Analysis*. Cambridge University Press, Cambridge, UK, 2004.

[TSK06]   P.-N. Tan, M. Steinbach, and V. Kumar. *Introduction to Data Mining*. Pearson Addison Wesley, Boston, 2006.

[WH06]   Y. Wang and J. Hodges. Document clustering with semantic analysis. In *Proceedings of the Hawaii International Conference on System Sciences (HICSS'06)*, IEEE, Los Alamitos, CA, 2006.

[WZW85]   S.K. Michael Wong, W. Ziarko, and P.C.N. Wong. Generalized vector space model in information retrieval. In *Proceedings of the ACM SIGIR Conference on Research and Development in Information Retrieval*, pages 18–25, ACM Press, New York, 1985.

## Part II

Document Retrieval and Representation

6

Vector Space Models for Search and Cluster Mining

Mei Kobayashi and Masaki Aono

Overview

This chapter reviews some search and cluster mining algorithms based on vector space modeling (VSM). The first part of the review considers two methods to address polysemy and synonomy problems in very large data sets: *latent semantic indexing* (LSI) and *principal component analysis* (PCA). The second part focuses on methods for finding minor clusters. Until recently, the study of minor clusters has been relatively neglected, even though they may represent rare but significant types of events or special types of customers. A novel new algorithm for finding minor clusters is introduced. It addresses some difficult issues in database analysis, such as accommodation of cluster overlap, automatic labeling of clusters based on their document contents, and user-controlled trade-off between speed of computation and quality of results. Implementation studies with new articles from Reuters and *Los Angeles Times* TREC datasets show the effectiveness of the algorithm compared to previous methods.

6.1 Introduction

Public and private institutions are being overwhelmed with processing information in massive databases [WF99]. Since the documents are generated by different people or by machines, they are of heterogeneous format and may contain different types of multimedia data (audio, image, video, HTML) and text in many different languages. A number of successful methods for retrieving nuggets of information have been developed by researchers in the data mining community.[1] This chapter examines vector space modeling (VSM), an effective tool for information retrieval (IR) [BYRN99] introduced by Salton [Sal71] over three decades ago. It reviews methods for enhancing the scalability of VSM to enable mining information from large databases on the order of magnitude greater than originally envisioned by Salton. Several features of VSM make an attractive method:

---

[1] *www.kdnuggets.com*

- It can handle documents of heterogeneous format.
- It can handle different types of multimedia data.
- It can facilitate processing of documents in many different languages.
- Query, search, retrieval, and ranking can be fully automated.
- Most of the computational workload is carried out during the preprocessing stage, so query and retrieval are relatively fast.

The second part of this chapter examines methods based on VSM for finding both *major* and *minor* clusters.[2] In real-world applications, topics in major clusters are often known to practitioners in the field, for example, through interactions with customers and observations of market trends. In contrast, information in minor clusters cannot be discerned from daily experience, and until recently, demand for methods to facilitate their discovery and analysis has been relatively weak. Although successful techniques have been developed for identifying major clusters, few techniques have been developed for understanding smaller, *minor* clusters. However, the landscape is changing.

Recently, corporate, government, and military planners are recognizing that mining even a portion of the information in minor clusters can be extremely valuable [SY02]. For example:

- Corporations may want to mine customer data to find minor reasons for customer dissatisfaction (in addition to the major reasons), since minor clusters may represent emerging trends or long-term, small dissatisfactions that may lead users to switch to another product.
- Credit card and insurance firms may want to better understand customer data to set interest and insurance rates.
- Security agencies may want to use mining technologies for profile analysis.
- Scientists may want to mine weather and geographical data to refine their forecasts and predictions of natural disasters.

This chapter is organized as follows. Section 6.2 reviews basic terminology and mathematical tools used in VSM, then introduces some methods for increasing the scalability of IR systems based on VSM. Section 6.3 reviews clustering, another approach for addressing the scalability problem associated with VSMs. In particular, it examines nonpartitioning approaches for mining clusters. Section 6.4 introduces an algorithm for mining major and minor clusters, and then examines variations that can significantly reduce computational and memory requirements, with only a slight decrease in the number of retrieved clusters. Users may select the best variation for their application based on their own cost-benefit analyses. Results from implementation studies with the algorithm and its variations are presented in Section 6.5. This chapter concludes with a discussion of open problems and possible directions for future research.

---

[2] *Major* and *minor* clusters are are relatively *large* and relatively *small* clusters with respect to a database under consideration.

6.2 Vector Space Modeling (VSM)

6.2.1 The Basic VSM Model for IR

VSM has become a standard tool in IR systems since its introduction over three decades ago [BDJ99, BDO95, Sal71]. One of the advantages of the method is its support of relevance ranking of documents of heterogeneous format with respect to queries. The attributes must be well-defined characteristics of the documents, such as keywords, key phrases, names, and time stamps.

In *Boolean* vector models, each coordinate of a document vector is naught (when the corresponding attribute is absent) or unity (when the corresponding attribute is present). *Term weighting* is a common refinement of Boolean models that takes into account the frequency of appearance of attributes (such as keywords and key phrases), the total frequency of appearance of each attribute in the document set, and the location of appearance (for example, in the title, section header, abstract, or text).

A fairly common type of term weighting is *term frequency inverse document frequency weighting (tf-idf)*, in which the weight of the ith term in the jth document, denoted by weight(i, j), is defined by

$$\text{weight}(i, j) = \begin{cases} (1 + tf_{i,j}) \log_2(N/df_i), & \text{if } tf_{i,j} \geq 1, \\ 0, & \text{if } tf_{i,j} = 0, \end{cases}$$

where $tf_{i,j}$ is the number of occurrences of the ith term within the jth document d_j, and df_i is the number of documents in which the term appears [MS00]. There are many variations of the *tf-idf* formula. All are based on the idea that term weighting reflects the importance of a term within a given document and within the entire document collection. *Tf-idf* models assume that the *importance* of a term in a document is reflected by its *frequency of appearance* in documents.[3]

In IR systems, queries are modeled as vectors using the same attribute space as the documents. The relevancy ranking of a document with respect to a query depends on its *similarity distance* to the query vector. Many systems use the cosine of the angle defined by the query and document vectors to "*measure*" the similarity, because it is relatively simple to compute,[4] and implementation experiments have indicated that it tends to lead to better IR results than the Euclidean distance (Figure 6.1). Numerous other distances have been proposed for various applications and database sizes [Har99].

---

[3] Some researchers believe that this fundamental assumption of correlation between term frequency and its importance is not valid. For example, when a document is about a single, main subject, it does not need to be explicitly mentioned in every sentence since it is tacitly understood. Consequently, its term frequency may be quite low relative to its importance. However, if a document is about multiple subjects, each time the subject (person or object) changes, it must be explicitly mentioned to avoid ambiguity, so the term frequencies of all subjects will be very high [Kat96].

[4] The cosine of the angle defined by two normalized vectors is their inner product. Since document vectors are very sparse (usually less than 2% nonzero), the computation is fast and simple.

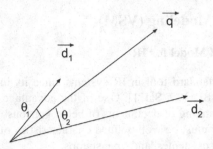

Fig. 6.1. Similarity ranking of documents d_1 and d_2 with respect to query q. The first document d_1 is "*closer*" to the query q when the distance is defined as the cosine of the angle made between the corresponding vectors. The second document d_2 is "*closer*" to the query q in the Euclidean norm. That is, $\cos \theta_1 < \cos \theta_2$, while $\|d_1 - q\|_2 > \|d_2 - q\|_2$.

Many databases are so massive that the inner product-based ranking method requires too many computations and comparisons for real-time response. Scalability of relevancy ranking methods is a serious concern as users consistently select the most important feature of IR engines to be a fast, real-time response to their queries.[5] One approach toward solving this problem is to reduce the dimension of mathematical models by projection into a subspace of sufficiently small dimension to enable fast response times, but large enough to retain characteristics for distinguishing contents of individual documents. This section reviews and compares two algorithms for carrying out dimensional reduction: *latent semantic indexing* (LSI) and a variation of *principal component analysis* (PCA). Some other approaches for reducing the dimension of VSM-based IR are centroid and least squares analysis [PJR01] and a Krylov subspace method [BR01].

6.2.2 Latent Semantic Indexing (LSI)

Given a database with M documents and N distinguishing attributes for relevancy ranking, let A denote the corresponding M-by-N document-attribute matrix model with entries $a(i, j)$ that represent the importance of the ith term in the jth document. The fundamental idea in LSI is to reduce the dimension of the IR problem to k, where $k \ll M, N$, by projecting the problem into the space spanned by the rows of the closest rank-k matrix to A in the Frobenius norm [DDF+90]. Projection is performed by computing the singular value decomposition (SVD) of A [GV96], and then constructing a modified matrix A_k, from the k largest singular values σ_i ; $i = 1, 2, \ldots, k$, and their corresponding vectors:

$$A_k = U_k \Sigma_k V_k^T .$$

Σ_k is a diagonal matrix with monotonically decreasing diagonal elements σ_i. The columns of matrices U_k and V_k are the left and right singular vectors of the k largest singular values of A.

---

[5] Graphics, Visualization, and Usability Center of Georgia Institute of Technology (GVU) Web users' survey: www.gvu.gaetch.edu/user_surveys

Processing the query takes place in two steps: projection, followed by matching. In the projection step, input queries are mapped to pseudo-documents in the reduced query-document space by the matrix U_k, then weighted by the corresponding singular values σ_i from the reduced rank, singular matrix Σ_k:

$$q \longrightarrow \hat{q} = q^T U_k \Sigma_k^{-1},$$

where q represents the original query vector, $\hat{q}$ the pseudo-document, q^T the transpose of q, and $(\cdot)^{-1}$ the inverse operator. In the second step, similarities between the pseudo-document $\hat{q}$ and documents in the reduced term document space V_k^T are computed using a similarity measure, such as angle defined by a document and query vector.

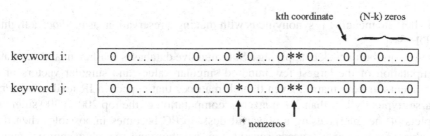

Fig. 6.2. LSI handles synonymy. Keywords i and j are synonyms that were inadvertently selected as distinct attributes during vector space modeling. When the dimension is reduced from N to k in LSI, the ith and jth keyword vectors are mapped to similar vectors, because the synonyms appear in similar contexts.

The inventors of LSI claim that the dimensional reduction process reduces unwanted information or *noise* in the database, and aids in overcoming the synonymy and polysemy problems. *Synonymy* refers to the existence of equivalent or similar terms that can be used to express an idea or object in most languages, while *polysemy* refers to the fact that some words have multiple, unrelated meanings [DDF+90]. For example, LSI correctly processes the synonyms *car* and *automobile* that appear in the Reuters news article set by exploiting the concept of *co-occurrence of terms*. Synonyms tend to appear in documents with many of the same terms (i.e., text document attributes), because the documents will tend to discuss similar or related concepts. During the dimensional reduction process, co-occurring terms are projected onto the same dimension (Figure 6.2). Failure to account for synonymy will lead to many small, disjoint clusters, some of which should be clustered together, while failure to account for polysemy can lead to clustering unrelated documents together.

Conversely, a word with multiple, unrelated meanings has many contexts in which it appears. The word will be mapped to a different region for each separate meaning and context because the existence of co-occurring terms is unlikely for documents that cover different meanings of a word (Figure 6.3). An example of polysemy is the word *book*, which may be synonymous with novel, biography, or text.

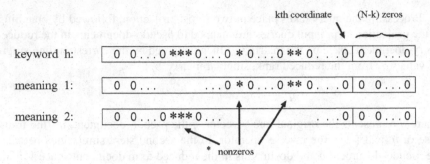

Fig. 6.3. LSI handles polysemy. Words with multiple meanings are mapped to distinct areas in the LSI subspace for each meaning.

An alternate meaning is synonymous with making a reservation, as in "*book* a flight" [DDF+90].

A major bottleneck in applying LSI to massive databases is efficient and accurate computation of the largest few hundred singular values and singular vectors of a document-attribute matrix. Even though matrices that appear in IR tend to be very sparse (typically less than 2% nonzero), computation of the top 200 to 300 singular triplets of the matrix using a powerful desktop PC becomes impossible when the number of documents exceeds several hundred thousand. An algorithm presented in the next section, overcomes some scalability issues associated with LSI, while handling synonymy and polysemy in an analogous manner.

6.2.3 Principal Component Analysis (PCA)

The scalability issue can be resolved while effectively handling synonymy and polysemy by applying a dimensional reduction method known as *principal component analysis* (PCA) [Jol02]. Invented first by Pearson [Pea01] in 1901 and independently reinvented by Hotelling [Hot33] in 1933, PCA has several different names such as the *Kahrhunen-Loève procedure*, *eigenvector analysis*, and *empirical orthogonal functions*, depending on the context in which one is being used. Until recently it has been used primarily in statistical data analysis and image processing.

We review a PCA-based algorithm for text and data mining that focuses on *covariance matrix analysis* (COV). In the COV algorithm, document and query vectors are projected onto the subspace spanned by the k eigenvectors for the largest k eigenvalues of the covariance matrix of the document vectors C, that is, the IR problem is mapped to a subspace spanned by top k principal components. Stated more rigorously, given a very large database modeled by a set of M document vectors $\{d_i^T \mid i = 1, 2, \ldots, M\}$ with N attributes, the associated document *covariance matrix* is

$$C \equiv \frac{1}{M} \sum_{i=1}^{M} d_i d_i^T - \bar{d}\,\bar{d}^T \,,$$

where d_i represents the ith document vector and $\bar{d}$ is the component-wise average over the set of all document vectors [MKB79]. Since the covariance matrix is symmetric, positive semidefinite, it can be decomposed into the product

$$C = V \, \Sigma \, V^T .$$

Here V is an orthogonal matrix that diagonalizes C so that the diagonal entries of Σ are in monotone decreasing order going from top to bottom, that is, diag(Σ) = $(\lambda_1, \lambda_2, \ldots, \lambda_N)$, where $\lambda_i \geq \lambda_{i+1}$ for $i = 1, 2, \ldots, N - 1$ [Par97].

To reduce the dimension of the IR problem to $k \ll M, N$, project all document vectors and the query vector into the subspace spanned by the k eigenvectors $\{v_1, v_2, \ldots, v_k\}$ corresponding to the largest k eigenvalues $\{\lambda_1, \lambda_2, \ldots, \lambda_k\}$ of the covariance matrix C. Relevancy ranking with respect to the modified query and document vectors is performed in a manner analogous to the LSI algorithm, that is, projection of the query vector into the k dimensional subspace followed by measurement of the similarity.

6.2.4 Comparison of LSI and COV

Implementation studies for IR show that LSI and COV are similar. Both project a high-dimensional problem into a subspace to reduce the computational cost of ranking documents with respect to a query, but large enough to retain distinguishing characteristics of documents to enable accurate retrieval [KAST02].

However, the algorithms differ in several ways. For instance, they use different criteria to determine a subspace: LSI finds the space spanned by the rows of the closest rank-k matrix to A in the Frobenius norm [EY39], while COV finds the k-dimensional subspace that best represents the full data with respect to the minimum square error. COV shifts the origin of the coordinate system to the "center" of the subspace to spread apart documents to better distinguish them from one another.

A second advantage of the COV algorithm is scalability. The primary bottleneck of COV is the computation of the largest few hundred eigenvalues and corresponding eigenvectors of a square, symmetric, positive semidefinite matrix with height and width less than or equal to the dimension of the attribute space. Because the dimension of a covariance matrix is independent of the number of documents, COV can be used for real-time IR and data mining as long as the attribute space is relatively small. Usually the dimension of the attribute space is less than 20,000, so computations can be performed in the main memory of a standard PC. When the dimension of the attribute space is so large that the covariance matrix cannot fit into the main memory, the eigenvalues of the covariance matrix can be computed using an implicit construction of the covariance matrix [MKS00]. Since paging will occur, this algorithm is much slower, especially for databases with hundreds of thousands of documents.

An alternative method suitable for limited applications is neural networks for estimation of the eigenvalues and eigenvectors of the covariance matrix [Hay99]. The advantage of the neural network approach is a reduction in computational expense.

However, the disadvantages make the method unattractive for IR. For example, it is not clear when convergence to an eigenvector has occurred (although convergence can be tested, to some extent, by multiplying the vector to the matrix and examining the difference in the angle made by the input and output vectors). Also, it is not clear which eigenvector has been found, that is, whether it is the ith or $(i+1)$st eigenvalue. And sometimes the neural network is not converging toward anything at all. In short, neural network approaches are not suitable for IR and clustering applications since there is no guaranteed means by which the eigenvectors corresponding to the largest k eigenvalues can be computed.

A third attractive feature of COV is an algorithm for mining information from datasets that are distributed across multiple locations [QOSG02]. The main idea of the algorithm is to compute *local* principal components for dimensional reduction for each location. Information about local principal components is subsequently sent to a centralized location, and used to compute estimates for the *global* principal components. The advantage of this method over a centralized (nondistributed) approach and parallel processing approach is the savings in data transmission rates. Data transmission costs often exceed computational costs for large datasets [Dem97]. More specifically, transmission rates will be of order $O(sp)$ instead of $O(np)$, where n is the number of all documents over all locations, p is the number of attributes, and s is the number of locations. According to the authors, when the dominant principal components provide a good representation of the datasets, the algorithm can be as equally accurate as its centralized (nondistributed) counterpart in implementation experiments. However, if the dominant components do not provide a good representation, up to 33% more components need to be computed to attain a level of accuracy comparable to its centralized counterpart, and then the subspace into which the IR problem will be mapped will be significantly larger.

6.3 Clustering Methods

Cluster analysis can also be used to understand topics addressed by documents in massive databases [Ham03]. It is important to note that unlike an arithmetic computation, there is no "correct solution" to the clustering problem for very large databases. Depending on the perspective, two documents may or may not be similar. Studies have shown that manual clustering and human evaluation of clustering results are riddled with inconsistencies [MBDH98]. The work provides insight into the difficulty of automating this task. Nevertheless, studies have also shown that despite imperfections associated with automated and human clustering methods, they may still provide valuable insights into the contents of databases.

This section focuses on clustering methods and how they handle cluster overlaps, that is, whether or not documents are allowed to belong to more than one cluster. From this perspective, there are at least four categories of methods:

- *Hard clustering methods* that require each document to belong to *exactly one* cluster;

- *Hard clustering methods* that permit each document to belong to *at most one* cluster (documents can be classified as an outlier or noise, in which case they do not belong to any cluster);
- *Soft clustering methods* that accommodate overlapping clusters, although they do not actively target the discovery of overlapping clusters; and
- *Soft clustering methods* that *actively* support mining of overlapping clusters and overlap information.

Many well-known, classic clustering algorithms are partition-based, hard clustering methods. Examples include *k-means*, *k-medoid*, *k-modes*, and *k-prototypes* algorithms; *agglomerative* and *divisive algorithms*; and *gravitational methods* [HK00]. Variations of these hard clustering methods have been developed in which appropriate documents can be classified as noise rather than be forced to be associated with a cluster.

Soft clustering methods have been less publicized. However, they are equally important. Some classic methods that are not scalable to larger databases are reviewed in [ELL01]. Methods that use fractional assignment of documents to more than one cluster tend to emphasize the computation of the assignments rather than understanding the contents of the database through mining information about relationships between the overlapping clusters.

More recent works on soft clustering are designed to process document collections that are many orders larger. Overlapping cluster analysis and good visualization help users understand situations in which many categories are associated with a set of documents: "A one-document/one-category assumption can be insufficient for guiding a user through hundreds or thousands of articles" [Hea99]. Works from computer science [KG99], machine learning [Bez81, PM01], and Web page mining [LK01, ZE98] perspectives also advocate membership in multiple clusters with different degrees of association.

Houle [Hou03] proposed a soft clustering method that recognizes the importance of mining clusters overlaps, and actively looks for and accommodates their presence. His method is based on use of a new and efficient type of data structure for fast, approximate nearest-neighbor search. It may be implemented using VSM.

Kobayashi and Aono [KA06] proposed a soft clustering algorithm that differs from those listed above in several respects. It appears to be the only soft clustering algorithm for massive (non-Web) document databases that handles the synonymy and polysemy problems to some extent and reduces the dimensionality of the problem through projections, random sampling, and a user-specified trade-off between the speed and accuracy of the results.

6.3.1 Minor Cluster Mining–Related Work

Implementation studies show that LSI and COV can successfully find *major* document clusters [KA02]. However, they are not as successful at finding smaller, *minor* clusters, because major clusters dominate the process. During dimensional reduction in LSI and COV, documents in minor clusters are often mistaken for noise or placed arbitrarily into any cluster.

Ando [And00] proposed *iterative rescaling*, an algorithm for clustering small collections of documents (such as personal email archives) and conducted implementation studies with a set of 683 articles from TREC.[6] The main idea of the algorithm is to prevent major themes from dominating the selection of basis vectors in the subspace into which the clustering and IR problem will be mapped. A fixed, unfavorable bias is introduced to documents that belong to clusters that are well represented by basis vectors that have already been selected. The weight for unfavorable biasing is based on the *magnitude* (length in the Euclidean norm) of the *residual* of each document vector.[7]

The algorithm is somewhat successful in detecting clusters; however, the following problems can occur when the number of documents is large (greater than a few thousand): fixed weighting can obscure the associations of documents that belong to more than one cluster after one of the clusters to which it belongs is identified; all minor clusters may not be identified; the procedure for finding eigenvectors may become unstable when the scaling factor q is large; the basis vectors b_i are not always orthogonal; and if the number of documents in the database is very large, the eigenvector cannot be computed on an ordinary PC, because the residual matrix becomes dense after only a few iterations, leading to a memory overflow.

Recently, Kobayashi and Aono [KA02] proposed two algorithms for identifying (possibly overlapping) multiple major and minor document clusters that overcome some of the difficulties associated with the iterative rescaling algorithm. Their primary new contribution, called *LSI with rescaling* (*LSI-RS*), is the introduction of dynamic control of the weighting to reduce loss of information about minor clusters. A second modification replaces the computation of eigenvectors in the iterative rescaling algorithm with the computation of the SVD for robustness. A third modification is the introduction of modified Gram-Schmidt orthogonalization of the basis vectors [GV96].

A second algorithm proposed in [KA02] for minor cluster identification, called *COV-rescale* (*COV-RS*), is a modification of COV, analogous to LSI-RS and LSI. The COV-RS algorithm computes the residual of the covariance matrix. Our implementation studies indicate that COV-RS is better than LSI, COV, and LSI-RS at identifying large and multiple minor clusters.

Both LSI-RS and COV-RS are computationally expensive, because they require rescaling of all document vectors before computation of each additional basis vector for subspace projection. For even moderately large databases, after a few iterations the document attribute matrix becomes dense so that main memory constraints become a bottleneck. Details on a comparison study of LSI-RS, COV-RS, and iterative rescaling algorithms are available in [KA02]. In the next section we propose techniques for enhancing the LSI-RS algorithm to overcome the scalability issue for very large databases.

---

[6] Text REtrieval Competition (TREC) sponsored by the United States National Institute of Standards and Technology (NIST): *http://trec.nist.gov*

[7] The residual of a document vector is the proportion of the vector that cannot be represented by the basis vectors that have been selected thus far.

6.4 Selective Scaling, Sampling, and Sparsity Preservation

The basic COV algorithm is recommended for mining major clusters. This section introduces *COV with selective scaling* (COV-SS) [KA06], a more efficient minor cluster mining algorithm than COV-RS [KA02]. Like COV-RS, COV-SS allows users to skip over iterations that find repeats of major clusters and jump to iterations to find minor clusters. In addition, COV-SS reduces computational, storage, and memory costs by quickly testing whether rescaling is necessary. Although the savings per iteration is modest, the total savings is large because sparsity is preserved over several or more subsequent iterations. When random sampling is incorporated into the COV-SS algorithm, it will increase minor cluster mining capabilities to databases that are several orders of magnitude higher. To further increase the speed and scalability of the algorithm, modifications to COV-SS may be introduced to preserve the sparsity of the document vectors, at the expense of a slight degradation in the mining results. Empirical observations by the authors indicate that there may be at most, a few rare cases in which introduction of the perturbations may lead to poor results. No cases have been reported in practice [KA06]. Since the computational reduction is so significant, the sparsity preserving versions of the algorithm seem preferable and advantageous for most commercial applications.

The input parameters for algorithm COV-SS are denoted as follows. A is an M-by-N document attribute matrix for a dataset under consideration. k is the dimension to which the VSM will be reduced ($k \ll M, N$). ρ is a threshold parameter. And μ is the scaling offset. Initially, the residual matrix R is set to be A. Matrix R does not need to be kept in the main memory. It suffices to keep just the N-dimensional residual vector r_i during each of the M loops. The output is the set of basis vectors $\{b_i : i = 1, 2, \ldots, k\}$ for the k-dimensional subspace. M is either the total number of documents in the database or the number of randomly sampled documents from a very large database.

Here P and Q are M-dimensional vectors; R is the residual matrix (which exists in theory, but is not allocated in practice); r_i is the ith document vector of R (an N-dimensional vector); $\bar{r}$ is the component-wise average of the set of all residual vectors r, that is, $\bar{r} = (1/M) \sum_{i=1}^{M} \bar{r}$; C is the N-by-N square covariance matrix; w and t are double-precision floating point numbers; and *first* is a Boolean expression initially set equivalent to *true*.

COV-SS selectively scales document residual vectors based on the similarity measure $P[i]$ of the dot product of the most recently computed basis vector and the document vector. The user-specified threshold ρ and offset parameter μ control the number of minor clusters that will be associated with each basis vector. A small threshold and large offset value tend to lead to basis vectors associated with a large number of minor clusters. Conversely, a large threshold and a small offset value tend to lead to basis vectors with few minor clusters.

The computational work associated with the COV-based rescaling algorithms (COV-RS and COV-SS) is significantly greater than the basic COV algorithm. Rescaling is computationally expensive for large databases. COV has no rescaling costs, but it uses a moderately expensive eigenvalue-eigenvector solver for large, symmetric

---

Algorithm 6.4.1 COV-SS: $(\mathbf{A}, k, \rho, \mu, \mathbf{b})$

---

for (int $h = 1, h \leq k, h++$) {
  if (! first) **for** (int $i = 1, i \leq M, i++$) {
    $t = |\mathbf{r}_i|$;   (length of document vector)
    if ($\| \mathbf{P[i]} \| > \rho$) {   (dot product greater than threshold)
      $w = (1 - \|\mathbf{P[i]}\|)^{(t+\mu)}$;   (compute scaling factor)
      $\mathbf{r}_i = \mathbf{r}_i w$;   (selective scaling)
      continue;
    }
  }
  $\mathbf{C} = (1/M) \sum_{i=1}^{M} \mathbf{r}_i \mathbf{r}_i^T - \bar{\mathbf{r}} \bar{\mathbf{r}}^T$;   (compute covariance matrix)
  $\mathbf{b}_h$ = PowerMethod($\mathbf{C}$) ;   (compute λ_{max} and its eigenvector)
  $\mathbf{b}_h$ = MGS($\mathbf{b}_h$) ;   (Modified Gram-Schmidt)
  for (int $i = 1, i \leq M, i++$) {
    $\mathbf{Q[i]} = \mathbf{r}_i \cdot \mathbf{b}_h$; $\mathbf{P[i]} = \|\mathbf{r}_i\|^2$;
    $\mathbf{P[i]} = \mathbf{Q[i]} / \sqrt{\mathbf{P[i]}}$;   (store dot product = similarity measure)
  }
  for (int $i = 1, i \leq M, i++$) ; $\mathbf{r}_i = \mathbf{r}_i$ - $\mathbf{Q[i]} \mathbf{b}_h$;   (residual)
  if (first) first = false ;
}

---

positive semidefinite matrices. COV-RS and COV-SS use the accurate and inexpensive *power method* [GV96] to find the largest eigenvalue and its corresponding eigenvector after each round of (possible) rescaling of residual vectors. A *selective scaling algorithm for LSI*, which will be denoted by LSI-SS, can be constructed in an analogous manner. It suffers from the same difficulties associated with sparsity preservation as COV-SS.

Mining minor clusters using the selective scaling algorithm for COV becomes prohibitively expensive if the database under consideration is large. One way to overcome the problem is to introduce random sampling of documents. COV-SS is applied to the covariance matrix constructed from a set of randomly selected document vectors in the database. Since different results can be expected for different samples, this process of sampling followed by selective scaling should be repeated as many times as the stopping criterion permits. The stopping criterion may determined by a number of factors, such as computational, memory and storage resources or the number of new clusters found during each sampling. In setting the sample sizes, the user needs to recognize the computational trade-off between sample sizes and the number of times sampling must be performed to find most minor clusters. Larger samples increase the cost of selective scaling. However, smaller samples are likely to lead to identification of fewer clusters.

The major bottleneck in selective scaling is the conversion of sparse vectors into dense vectors. However, many of the nonzero coordinates are very small relative to the original nonzero coordinates. To further increase the speed and scalability of the algorithm, additional methods for preserving the sparsity of the document vectors may be employed. One idea is to select a threshold ϵ such that all nonzeros in the residual matrix smaller than ϵ will be reset to zero after rescaling. A larger ϵ threshold will lead to greater preservation of sparsity, and hence more reduction in computation and data access times, but it will also introduce larger numerical error during computation. Another method is to consider the very small nonzero coordinates of a newly computed basis vector. Before rescaling the residual matrix, set the small values in the basis vector to zero to reduce the computational cost of rescaling and the introduction of errors. The first ϵ method was tested in implementation studies that are described in the next section.

6.5 Implementation Studies

This section describes implementation studies with the LSI- and COV-based algorithms described in previous sections.

6.5.1 Studies with TREC Benchmark Data Sets

Numerical experiments were carried out with LSI, COV, COV-RS, and COV-SS with and without sampling using the Reuters and *Los Angeles Times* news databases from TREC (with 21,578 and 127,742 articles, respectively) [KA06]. Results from both the LSI- and COV-based search engines were good. In experiments with the Reuters and *Los Angeles Times* articles, the LSI and COV algorithms could be used to find major clusters, but they usually failed to find all minor clusters. The algorithms lost information in some minor clusters, because major clusters and their large subclusters dominate the subjects that will be preserved during dimensional reduction.

Table 6.1 summarizes the strengths and limitations of various cluster identification algorithms based on theoretical considerations and results from implementation studies. For medium-size databases, major clusters should be mined using basic COV and minor clusters using COV with selective scaling. Major cluster identification results from LSI and COV are usually identical. However, COV usually requires 20% to 30% fewer iterations to find major clusters because it can detect clusters in both the negative and positive directions along each basis vector, since the origin is shifted during the dimensional reduction phase in pre-computations. LSI can only detect clusters either in the positive direction or in the negative direction of each basis vector, but not both.

For massive databases, major clusters should be mined using basic COV (since LSI is not scalable to massive databases), and COV with selective scaling and sampling should be used to find minor clusters. Selective scaling is preferable to complete rescaling since the results from both should be similar, but selective scaling is more computationally efficient.

Figure 6.4 shows results from cluster identification experiments with basic COV and its variations for the *Los Angeles Times* dataset. It displays the number of major and minor clusters retrieved using COV, COV with rescaling and sampling, and COV with selective scaling and sampling with and without replacement of small nonzero entries by zero. The computational complexity of computing 20 basis vectors from

Table 6.1. LSI- and COV-based cluster analysis algorithms

Algorithm		Scalability		Clusters found	
	Variations	DB size	Speed	Major	Minor
LSI	basic	++	++	++++	+
	iterative RS	+	+	++	++
	RS	+	+	++++	+++
	SS	+	++	++++	+++
	sampling	+++	+++	+	+++
	RS + sampling	++	++	+	+++
	SS + sampling	++	+++	+	+++
COV	basic	+++	++	++++	+
	RS	+	+	++++	+++
	SS	+	+	++++	+++
	sampling	+++	++	+	+++
	RS + sampling	+++	++	+	+++
	SS + sampling	+++	+++	+	+++
	SS + sampling w/sparsity	++++	++++	+	++++

DB size: ++++ = very large, +++ = large, ++ = medium, + = small.
Speed: ++++ = fast, +++ = moderate, ++ = slow, + = prohibitive.
No. of clusters found: ++++ = most , +++ = many , ++ = some , + = few.

this dataset is summarized in Table 6.2. Several conclusions can be drawn from these results:

- COV with selective scaling and sampling finds the largest number of clusters. Straightforward COV and COV with rescaling and sampling find approximately the same number of clusters.
- Straightforward COV finds the largest number of major clusters, and its ratio of major to minor clusters is the largest among the three methods.
- COV with rescaling and sampling and COV with selective scaling and sampling find approximately the same number of major clusters. However, COV with selective scaling and sampling finds many more minor clusters (see Figure 6.5.)
- Overlaps between document clusters appear to be more common than isolated clusters. Our algorithms preserve overlaps that naturally occur in databases. Figure 6.6 shows overlapping clusters in a three-dimensional slice of document space. In some three-dimensional projected views, clusters overlap to a greater or lesser extent, or more than one pair may overlap.

6.6 Directions for Future Work

This chapter concludes by mentioning a number of interesting questions and open problems in information retrieval and clustering:

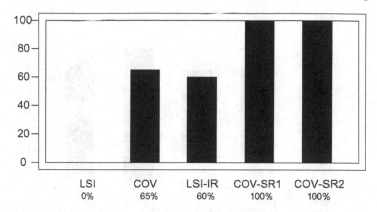

Fig. 6.4. Clusters (major and minor combined) retrieved using COV, COV with rescaling and sampling, and COV with selective scaling and sampling. Sampled data are averaged over three runs.

Table 6.2. Computation time and storage for determining 20 basis vectors of an artificial dataset (Figure 6.4)

	LSI	COV	LSI-IR	COV-SS+samp	COV-SS+samp+sparse
Time (sec)	0.56	1.8	104.5	1488.9	49.1
Memory (MB)	3	3	70	65	11

- *How can users determine the optimal dimension k to which the attribute space should be reduced?* Clearly, the answer will depend on the computational hardware resources, the dimension of the original attribute space, the number of polysemy and synonymy problems associated with the selection of the original attributes, and other factors. Some ideas for resolving this problem have been proposed recently by Dupret [Dup03] and Hundley and Kirby [HK03].
- *How can one determine whether a massive database is suitable for random sampling?* That is, are there simple tests to determine whether or not there is an abundance of witnesses or whether the database consists entirely of noise or documents on completely unrelated topics?
- *How can one devise a reliable means for estimating the optimal sampling size for a given database?* Factors to be considered are the cluster structure of the database (the number and sizes of the clusters and the amount of noise) and the trade-off between sample sizes and the number of samples. And are there any advantages to be gained from dynamically changing the sample sizes based on cluster that have already been identified?
- *What is a good stopping criterion for sampling?* That is, when is it appropriate for a user to decide that it is *likely* that *most* of the major or minor clusters have been found?
- *How can the GUI effectively map identified clusters and their interrelationships (subclusters of larger clusters, cluster overlap, etc.)?* Some successful

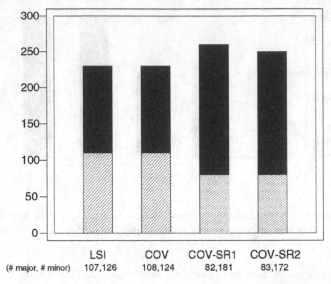

Fig. 6.5. Major clusters (light gray bar, bottom) and minor clusters (black bar, top) retrieved from the first 64 basis vectors computed for the *Los Angeles Times* news dataset using COV, COV with rescaling and sampling, and COV with selective scaling and sampling. Sampled data are averaged over three runs.

approaches for commercial applications have been reported by Ishii [Ish04] and Strehl [Str02].[8]

The study of methods for evaluating the effectiveness of a clustering algorithm or *cluster validation* is fast becoming an important area of research. Some techniques and validity indices are surveyed and proposed in [HBV01, NJT04, ZFLW02].

Acknowledgments

The authors thank Eric Brown of the IBM T.J. Watson Lab for helping us obtain access to TREC datasets, Michael E. Houle for technical discussions, Koichi Takeda of IBM Japan for his thoughtful management, and anonymous referees of journals and SIAM conferences for helpful suggestions. This work was conducted while both authors were members of the knowledge management team at IBM Tokyo Research Laboratory.

---

[8] Also see KDnuggets Software Visualization:
  http://www.kdnuggets.com/software/visualization.html

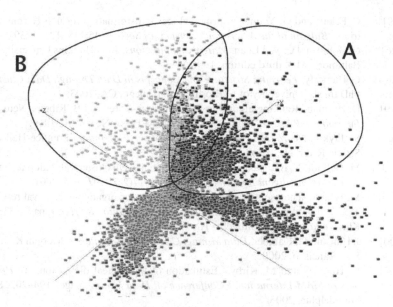

Fig. 6.6. Two overlapping document clusters on { *school council, tax, valley, California* } and { *school, art, music, film, child* } from the *Los Angeles Times* database. The clusters consist of 3,433 and 630 documents, respectively, and have 414 documents in common, with overlap ratios of 12.0% and 65.7%, respectively.

References

[And00]   R. Ando. Latent semantic space. In *Proceedings of ACM SIGIR*, pages 213–232. ACM Press, New York, 2000.

[BDJ99]   M.W. Berry, Z. Drmac, and E.R. Jessup. Matrices, vector spaces, and information retrieval. *SIAM Review*, 41(2):335–362, 1999.

[BDO95]   M.W. Berry, S.T. Dumais, and G.W. O'Brien. Using linear algebra for intelligent information retrieval. *SIAM Review*, 37(4):573–595, 1995.

[Bez81]   J. Bezdek. *Pattern Recognition with Fuzzy Objective Function Algorithms*. Plenum Press, New York, 1981.

[BR01]   K. Blom and A. Ruhe. Information retrieval using very short Krylov sequences. In *Proceedings of Computational Information Retrieval Workshop, North Carolina State University*, pages 3–24. SIAM, Philadelphia, 2001.

[BYRN99] R. Baeza-Yates and B. Ribeiro-Neto. *Modern Information Retrieval*. ACM Press, New York, 1999.

[DDF⁺90] S. Deerwester, S. Dumais, G. Furnas, T. Landauer, and R. Harshman. Indexing by latent semantic analysis. *Journal of the American Society for Information Science*, 41(6):391–407, 1990.

[Dem97]   J. Demmel. *Applied Numerical Linear Algebra*. SIAM, Philadelphia, 1997.

[Dup03]   G. Dupret. Latent concepts and the number of orthogonal factors in latent semantic analysis. In *Proceedings of ACM SIGIR*, pages 221–226. ACM Press, New York, 2003.

[ELL01]   B. Everitt, S. Landau, and N. Leese. *Cluster Analysis*. Arnold, London, UK, fourth edition, 2001.

[EY39]   C. Eckart and G. Young. A principal axis transformation for non-Hermitian matrices. *Bulletin of the American Mathematics Society*, 45:118–121, 1939.

[GV96]   G. Golub and C. Van Loan. *Matrix Computations*. John Hopkins University Press, Baltimore, MD, third edition, 1996.

[Ham03]   G. Hamerly. *Learning Structure and Concepts in Data Through Data Clustering*. PhD thesis, University of California at San Diego, CA, 2003.

[Har99]   D. Harman. Ranking algorithms. In R. Baeza-Yates and B. Ribeiro-Neto (eds.), *Information Retrieval*, pages 363–392, ACM Press, New York, 1999.

[Hay99]   S. Haykin. *Neural Networks: A Comprehensive Foundation*. Prentice-Hall, Upper Saddle River, NJ, second edition, 1999.

[HBV01]   M. Halkidi, Y. Batistakis, and M. Vazirgiannis. On cluster validation techniques. *Journal of Intelligent Infoormation Systems*, 17(2-3):107, 145, 2001.

[Hea99]   M. Hearst. The use of categories and clusters for organizing retrieval results. In T. Strzalkowski, editor, *Natural Language Information Retrieval*, pages 333–374. Kluwer Academic, Dordrecht, The Netherlands, 1999.

[HK00]   J. Han and M. Kamber. *Data Mining: Concepts & Techniques*. Morgan Kaufmann, San Francisco, 2000.

[HK03]   D. Hundley and M. Kirby. Estimation of topological dimension. In *Proceedings of SIAM International Conference on Data Mining*, pages 194–202. SIAM, Philadelphia, 2003.

[Hot33]   H. Hotelling. Analysis of a complex of statistical variables into principal components. *Journal of Educational Psychology*, 24:417–441, 1933.

[Hou03]   M. Houle. Navigating massive sets via local clustering. In *Proceedings of ACM KDD*, pages 547–552. ACM Press, New York, 2003.

[Ish04]   Y. Ishii. Analysis of customer data for targeted marketing: case studies using airline industry data (in Japanese). In *Proceedings of ACM SIGMOD of Japan Conference*, pages 37–49, 2004.

[Jol02]   I. Jolliffe. *Principal Component Analysis*. Springer, New York, second edition, 2002.

[KA02]   M. Kobayashi and M. Aono. Major and outlier cluster analysis using dynamic rescaling of document vectors. In *Proceedings of the SIAM Text Mining Workshop, Arlington, VA*, pages 103–113, SIAM, Philadelphia, 2002.

[KA06]   M. Kobayashi and M. Aono. Exploring overlapping clusters using dynamic rescaling and sampling. *Knowledge & Information Systems*, 10(3):295–313, 2006.

[KAST02]   M. Kobayashi, M. Aono, H. Samukawa, and H. Takeuchi. Matrix computations for information retrieval and major and outlier cluster detection. *Journal of Computational and Applied Mathematics*, 149(1):119–129, 2002.

[Kat96]   S. Katz. Distribution of context words and phrases in text and language modeling. *Natural Language Engineering*, 2(1):15–59, 1996.

[KG99]   S. Kumar and J. Ghosh. GAMLS: a generalized framework for associative modular learning systems. *Proceedings of Applications & Science of Computational Intelligence II*, pages 24–34, 1999.

[LK01]   K.-I. Lin and R. Kondadadi. A similarity-based soft clustering algorithm for documents. In *Proceedings of the International Conference on Database Systems for Advanced Applications*, pages 40–47. IEEE Computer Society, Los Alamitos, CA, 2001.

[MBDH98]   S. Macskassy, A. Banerjee, B. Davison, and H. Hirsh. Human performance on clustering Web pages. In *Proceedings of KDD*, pages 264–268. AAAI Press, Menlo Park, CA, 1998.

[MKB79]   K. Mardia, J. Kent, and J. Bibby. *Multivariate Analysis*. Academic Press, New York, 1979.

[MKS00]   L. Malassis, M. Kobayashi, and H. Samukawa. Statistical methods for search engines. Technical Report RT-5181, IBM Tokyo Research Laboratory, 2000.

[MS00]   C. Manning and H. Schütze. *Foundations of Statistical Natural Language Processing*. MIT Press, Cambridge, MA, 2000.

[NJT04]   Z.-Y. Niu, D.-H. Ji, and C.-L. Tan. Document clustering based on cluster validation. In *Proceedings of ACM CIKM*, pages 501–506. ACM Press, New York, 2004.

[Par97]   B. Parlett. *The Symmetric Eigenvalue Problem*. SIAM, Philadelphia, 1997.

[Pea01]   K. Pearson. On lines and planes of closest fit to systems of points in space. *The London, Edinburgh and Dublin Philosophical Magazine and Journal of Science, Sixth Series*, 2:559–572, 1901.

[PJR01]   H. Park, M. Jeon, and J. Rosen. Lower dimensional representation of text data in vector space based information retrieval. In M. Berry (ed.), *Proceedings of the Computational Information Retrieval Conference held at North Carolina State University, Raleigh, Oct. 22, 2000*, pages 3–24, SIAM, Philadelphia, 2001.

[PM01]   D. Pelleg and A. Moore. Mixtures of rectangles: interpretable soft clustering. In *Proceedings of ICML*, pages 401–408. Morgan Kaufmann, San Francisco, 2001.

[QOSG02] Y. Qu, G. Ostrouchov, N. Samatova, and A. Geist. Principal component analysis for dimension reduction in massive distributed datasets. In S. Parthasarathy, H. Kargupta, V. Kumar, D. Skillicorn, and M. Zaki (eds.), *SIAM Workshop on High Performance Data Mining*, pages 7–18, Arlington, VA, 2002.

[Sal71]   G. Salton. *The SMART Retrieval System*. Prentice-Hall, Englewood Cliffs, NJ, 1971.

[Str02]   A. Strehl. *Relationship-based clustering and cluster ensembles for high-dimensional data mining*. PhD thesis, University of Texas at Austin, 2002.

[SY02]   H. Sakano and K. Yamada. Horror story: the curse of dimensionality). *Information Processing Society of Japan (IPSJ) Magazine*, 43(5):562–567, 2002.

[WF99]   I. Witten and E. Frank. *Data Mining: Practical Machine Learning Tools and Techniques with Java Implementations*. Morgan Kaufmann, San Francisco, 1999.

[ZE98]   O. Zamir and O. Etzioni. Web document clustering: a feasibility demonstration. In *Proceedings of ACM SIGIR*, pages 46–54. ACM Press, New York, 1998.

[ZFLW02] O. Zaine, A. Foss, C.-H. Lee, and W. Wang. On data clustering analysis: scalability, constraints and validation. In *Proceedings of PAKDD, Lecture Notes in Artificial Intelligence, No. 2336*, pages 28–39. Springer, New York, 2002.

7

Applications of Semidefinite Programming in XML Document Classification

Zhonghang Xia, Guangming Xing, Houduo Qi, and Qi Li

Overview

Extensible Markup Language (XML) has been used as a standard format for data representation over the Internet. An XML document is usually organized by a set of textual data according to a predefined logical structure. It has been shown that storing documents having similar structures together can reduce the fragmentation problem and improve query efficiency. Unlike the flat text document, the XML document has no vectorial representation, which is required in most existing classification algorithms.

The kernel method and multidimensional scaling are two general methods to represent complex data in relatively simple manners. While the kernel method represents a group of XML documents by a set of pairwise similarities, the classical multidimensional scaling embeds data points into the Euclidean space. In both cases, the similarity matrix constructed by the data points should be semidefinite. The semidefniteness condition, however, may not hold due to the inference technique used in practice, which may lead to poor classification performance.

We will find a semidefinite matrix that is the closest to the distance matrix in the Euclidean space. Based on recent developments on strongly semismooth matrix valued functions, we solve the nearest semidefinite matrix problem with a Newton-type method.

7.1 Introduction

Along with dramatic growth of XML documents on the internet, it becomes more and more challenging for users to retrieve their desirable information. The demand of efficient search tools has led to extensive research in the area of information retrieval. XML documents are semistructured, including the logical structure and the textual data. Structural information plays an important role in information retrieval. A word may have different meanings depending on positions in a document. The semistructured data extracted from XML pages are usually maintained by relational database technologies [LCMY04], with which XML documents are decomposed and stored in corresponding tables. Studies [LCMY04] show that categorizing documents and storing their components according to their structures can reduce table fragmentation, and, thus, improve query performance.

Machine learning, building an automatic text classifier through learning pre-labeled documents, has become a standard paradigm in this field. Among many approaches [Seb02], support vector machines (SVMs) have achieved appealing performance in a variety of applications [Bur98, Joa98, ZI02, LCB+04]. Like other machine learning techniques, classical SVM algorithms are only applicable to vectorial data and not suitable for XML data containing structural information. A straightforward solution to vectorize a Web document involves the pairwise comparison of vector elements in which each element represents the similarity between a given document and another document. The number of dimensions of a vector, however, becomes huge, a so-called dimensional explosion problem when one uses this solution over a large set of documents.

Two possible solutions are the kernel method and multidimensional scaling (MDS). The kernel method [STV04] is a potential solution to cope with data having complex structures, for example, trees and graphs. Without vectorial formats of training data, a kernel-based learning algorithm requires only pairwise similarities among a set of data. These similarities are usually specified by a positive semidefinite matrix, called the *kernel matrix*. Along another direction, the MDS approach [CC01] embeds data points into a low-dimensional Euclidean space while preserving the original metric on the input data points if such a metric exists. A problem, however, arises when this method is applied to a set of XML documents without document type definitions (DTDs). When the DTDs of a set of XML documents are not provided, a common practice is to use some inference techniques (see [GGR+00]) to infer a schema from these sample documents. However, the edit distance estimated by inference techniques may not be Euclidean, resulting in deviation of the kernel matrix and MDS embedding.

In this work, we propose a novel method to compute a positive semidefinite matrix nearest to the estimated similarity matrix. The problem of searching a proper kernel matrix and a MDS embedding is formulated as semidefinite programming (SDP). Based on recent developments on strongly semismooth matrix valued functions, a Newton-type method is investigated to solve the SDP. We also report a new result characterizing the rank-deficient nearest semidefinite matrix, which is a property often desired in problems requiring a low-rank approximation. We test the proposed method on document sets selected from [DG] to justify this kernel method. Since existing inferred techniques can hardly work on large-size datasets, we modify the Xtract scheme to make it more efficient. Experimental studies show that our approach outperforms the algorithm with original similarity matrix in terms of the classification accuracy.

In the next section, we review some related work on XML document representation and classification. Section 7.3 introduces the background knowledge of the SVM classifier, the kernel method, MDS, and the measure of similarities among XML documents. In Section 7.4, we formulate the problem of finding a kernel matrix as SDP.

The Newton-type method is presented in Section 7.5. Experimental studies are presented in Section 7.6. Section 7.7 states the conclusion of this chapter.

7.2 Related Work

A vectorial representation method for an XML document has been proposed in [GJK+06] by using a vector of similarities between the document and a set of selected samples. Since an XML document can be represented by an ordered, labeled tree, tree edit distance is a common metric to measure the similarity/dissimilarity between two documents. Various algorithms have been proposed to compute the tree-edit distance between trees [SZ97, Che01, NJ02, CX05].

A structural rule-based classifier for XML documents is presented in [ZA03]. In the training phase, the classifier finds the structures that are most closely related to the class variable. In the test phase, the constructed rules are used to perform the structural classification. The proposed system is significantly more effective than an association-based classifier due to its ability of mining discriminatory structures in the data. [SZC+04] presented a method to optimize composite kernels for XML document classification. The paper showed that the problem of optimizing the linear combination of kernels is equivalent to a generalized eigenvalue problem, which leads to an efficient algorithm to construct the optimal composite kernel. Note that formulating an XML page as a linear combination of various media components (such as text, images) simplifies the construction of kernels since the kernel associated with each specific component is a positive semidefinite, and so is the linear combination. But this formulation ignores structural information of the XML data. A Bayesian network model is considered in [DG04] for semistructured document classification. The Bayesian model adopts the discriminant features via the Fisher kernel. Similar to [SZC+04], [DG04] also formulates an XML document as a linear combination of various media components, and does not fully utilize structural information of the XML data.

7.3 XML Document Representation and Classification

7.3.1 SVM Classifier

A binary setting of the classification problem can be described as follows. Given a set of training data $\{x_1, x_2, \ldots, x_l\}$ in some space R^p and their labels $\{y_1, y_2, \ldots, y_l\}$, where $y_i \in \{-1, +1\}$, estimate a prediction function f such that it can classify an unseen datapoint x.

The SVM learning algorithm aims to find a linear function of the form $f(x) = w^T x + b$, with $w \in R^p$ and $b \in R$ such that data point x is assigned to a label $+1$ if $f(x) > 0$, and a label -1 otherwise.

Let $\xi_i \geq 0, 1 \leq i \leq l$ be slack variables. Then, the linear SVM classifier can be obtained by solving the following optimization problem:

$$\min_{w,b,\xi} \tfrac{1}{2}(||w||)^2 + C\sum_{i=1}^{l} \xi_i$$
$$\text{s.t.} \quad y_i(w^T x_i + b) \geq 1 - \xi_i, \quad 1 \leq i \leq l,$$
$$\xi_i \geq 0, \qquad \qquad 1 \leq i \leq l,$$

where C is a predefined constant to control the trade-off between the gap of two classes and errors of classification.

7.3.2 Kernel Method

The kernel method represents data objects through a set of pairwise comparisons [STV04]. In the example of n XML documents given above, the dataset χ is represented by a $n \times n$ matrix, where the ijth element is a pairwise comparison $k_{ij} = k(x_i, x_j)$. Formally, a function $k : \chi \times \chi \longrightarrow R$ is called a positive definite kernel if and only if it is symmetric, that is, $k(x, x') = k(x', x)$ for any two objects $x, x' \in \chi$, and positive definite, that is,

$$\sum_{i=1}^{n}\sum_{j=1}^{n} c_i c_j k(x_i, x_j) \geq 0$$

for any $n > 0$, any choice of n objects $x_1, x_2, \ldots, x_n \in \chi$, and any choice of real numbers $c_1, \ldots, c_n \in R$.

It is well known that for any kernel function k on a space χ, there exists a Hilbert space $\mathcal{F}$ (also called the feature space) and a mapping $\phi : \chi \longrightarrow \mathcal{F}$ such that

$$k(x, x') = < \phi(x), \phi(x') >, \text{for any } x, x' \in \chi.$$

Hence, the similarity between x and x' is actually measured by the inner product of their images in the feature space $\mathcal{F}$. The kernel function k allows us to calculate the similarity without explicitly knowing the mapping ϕ. There are some well-known kernels, such as linear kernel, Gaussian RBF kernel, and sigmoid kernel. However, these kernels may not specifically adapt to XML documents.

7.3.3 MDS Method

Generally, for n data points $x_i, i = 1, \ldots, n$ with dissimilarities δ_{ij}, classic MDS attempts to find their coordinates in p dimensional space such that $\delta_{ij}^2 = (x_i - x_j)^T(x_i - x_j)$. These vectors can be written as a matrix $X = [x_1, x_2, \ldots, x_n]^T$. One first constructs a matrix $A = [a_{ij}]$ such that $a_{ij} = -\frac{1}{2}\delta_{ij}^2$. Then, matrix A is multiplied by centering matrix $H = I_n - \frac{1}{n}LL^T$ where $L = [1, 1, \ldots, 1]^T$, a vector of n ones. Let $B = HAH$. The following theorem [Bur05] states whether there exists a matrix X satisfying $B = XX^T$.

Theorem 1. *Consider the class of symmetric matrices A such that $a_{ij} > 0$, and $a_{ii} = 0$, for any i, j. Then B is positive semidefinite if and only if A is a distance matrix.*

Hence, if B is symmetric, positive semidefinite, one has a diagonal matrix Λ and matrix V whose columns are the eigenvectors of B such that $B = V\Lambda V^T$. If $p < n$, the rank of B is p and B has p nonnegative eigenvalues and $n - p$ zero eigenvalues. Let $\Lambda_p = diag(\lambda_1, \ldots, \lambda_p)$ and V_p be the corresponding eigenvector matrix. One has $X = V_p \Lambda_p^{\frac{1}{2}}$.

7.3.4 XML Documents and Their Similarities

An XML document can be represented as an ordered, labeled tree. Each node in the tree represents an XML element in the document and is labeled with the element tag name. Each edge in the tree represents the element nesting relationship in the document. Fig. 7.1. shows an example of a segment of an XML document and its corresponding tree.

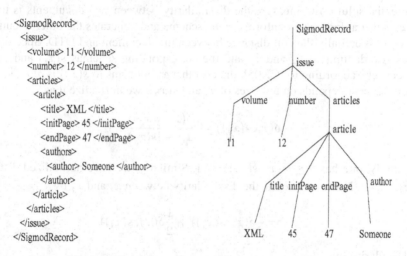

Fig. 7.1. SigmodRecord data and tree representation.

For the purpose of classification, a proper similarity/dissimilarity metric should be provided. Tree edit distance, which is a natural extension of string edit distance, can be used to measure the structural difference between two documents. Shasha and Zhang [SZ97] proposed three types of elementary editing operations for ordered, labeled forests: (1) insert, (2) delete, and (3) replace. Given two trees T_1 and T_2, the tree edit distance, denoted by $\delta(T_1, T_2)$, is defined as the minimum number of tree-edit operations to transform one tree to another.

However, using the tree edit distance between two documents directly may not be a good way of measuring the dissimilarity. Consider two SigmodRecord documents that contain 10 and 100 articles, respectively. To transform one tree to another, 90 insertions (or deletions) are required. Although they have similar structures, the tree edit distance can be very large.

```
<!ELEMENT SigmodRecord (issue)* >
<!ELEMENT issue (volume,number,articles) >
<!ELEMENT volume (#PCDATA)>
<!ELEMENT number (#PCDATA)>
<!ELEMENT articles (article)* >
<!ELEMENT article (title,initPage,endPage,authors) >
<!ELEMENT title (#PCDATA)>
<!ELEMENT initPage (#PCDATA)>
<!ELEMENT endPage (#PCDATA)>
<!ELEMENT authors (author)* >
<!ELEMENT author (#PCDATA)>
```

Fig. 7.2. An example of DTD for a SigmodRecord document.

Note that these documents may conform to the same DTD, meaning that they can be generated by the DTD. As an example, a SigmodRecord DTD is given in Fig. 7.2. DTD has been widely used to specify the schema of an XML document since it provides a simple way to specify the structure of an XML document. Hence, a potential solution to measure the dissimilarity between two documents is to use the cost that a document conforms to the schema and generates the other document. This cost is actually the edit distance between the document and DTD. Specifically, given two documents x_i and x_j and the corresponding schemas $s(x_i)$ and $s(x_j)$, respectively, according to [CX05], the cost that x_i confirms to $s(x_j)$ is $\delta(x_i, s(x_j))$. Since this cost depends on the sizes of x_i and $s(x_j)$, we normalize it as

$$\hat{\delta}(x_i, s(x_j)) = \frac{\delta(x_i, s(x_j))}{|x_i| + |s(x_j)|}.$$

Obviously, one has $0 \leq \hat{\delta}(x_i, s(x_j)) \leq 1$. Similarly, we have normalized distance $\hat{\delta}(x_j, s(x_i))$. Now, let's define the dissimilarity between x_i and x_j by

$$\delta_{ij} = \frac{1}{2}\hat{\delta}(x_i, s(x_j)) + \frac{1}{2}\hat{\delta}(x_j, s(x_i))$$

and similarity by

$$s_{ij} = 1 - \delta_{ij}. \tag{7.1}$$

However, not all XML documents provide DTDs in practice. In this case, to measure the similarity among documents, the inference technique [GGR$^+$00] has to be used to infer DTD schemas from a set of sample documents. That is, given a collection of XML documents, find a schema s, such that these document instances can be generated by schema s.

A schema can be represented by a tree in which edges are labeled with the cardinality of the elements. As a DTD may be recursive, some nodes may lead to a infinite path. A normalized regular hedge grammar [Mur00] became a common practice rather than DTD itself. Note that the definition of an element in a schema is independent of the definitions of other elements and it restricts only the sequence of subelements (the attributes are omitted in this work) nested within the element. Extracting a schema can be simplified by inferring a regular expression R (right linear grammar or nondeterministic finite automata) from a collection of input sequences

I satisfying (1) R is concise, that is, the inferred expression is simple and small in size; (2) R is general enough to accept sequences that are similar to those in I and (3) R is precise enough to exclude sequences not similar to those in I.

Inferring regular expressions from a set of strings has been studied in the Xtract scheme [GGR+00], which introduces minimum length description (MLD) to rank candidate expressions. In general, the MLD principle states that the best grammar (schema) to infer from a set of data is the one that minimizes the sum of the length of the grammar L_g and the length of the data L_d when encoded with the grammar. The overall goal is to minimize $L = L_g + L_d$.

According to our experimental studies, the Xtract scheme is not efficient for the large set of documents. To solve this problem, we modify the Xtract scheme, and the details will be introduced in the experimental studies when a similarity matrix is constructed.

7.3.5 Representation of XML Documents

To classify XML documents with SVM classifiers, the first thing to be addressed is how to represent the XML document as a standard input to SVM classifier. A general solution to represent XML documents can be described as follows. Define a mapping $\phi(x) : \chi \longrightarrow F = R^p$, where F is a Hilbert space, also called a feature space, and p is the dimension of the feature space. This means, for each document $x \in \chi$, its representation can be a vector of reals in R^p. One can choose a finite set of documents $x_1, x_2, \ldots, x_p \in \chi$ [STV04]. Then, any document $x \in \chi$ can be represented by a vector in the feature space:

$$x \in \chi \longmapsto \phi(x) = (s(x, x_1), s(x, x_2), \ldots, s(x, x_p))^T \in R^p,$$

where $s(x, y)$ is the similarity/dissimilarity between document x and y. The question is whether this vectorization is Euclidean. In other words, does the Euclidean distance between point x_i and x_j, denoted by d_{ij}, match the original dissimilarity δ_{ij}?

Since the schema is obtained by using an inference technique, a document may not exactly conform to its inferred schema. Note that a metric space is a pair (X, ρ), where X is a set and $\rho : X \times X \to [0, \infty)$ is a metric if it satisfies the following three axioms:

1. $\rho(u, v) = 0$ if and only if $u = v$,
2. $\rho(u, v) = \rho(v, u)$,
3. $\rho(u, v) + \rho(v, w) \geq \rho(u, w)$.

Unfortunately, the dissimilarity δ may not satisfy the third condition (i.e., the triangle inequality).

7.4 SDP Problem Formulation

We solve the problem by finding a positive semidefinite matrix nearest to the estimated distance matrix. Formally, given a symmetric matrix G, find a positive semidefinite matrix K such that the following norm is minimized:

$$
\begin{aligned}
&\min_{K \succeq 0} \|G - K\|^2 \\
&:= \min_{K \succeq 0} \sum_{i,j} (d_{ij} - k_{ij})^2,
\end{aligned}
\tag{7.2}
$$

where $K \succeq 0$ means that K is a positive semidefinite symmetric matrix and the norm is the Frobenius norm.

It is considered effective to normalize K so that all the diagonal elements are 1. Note that there is no close-form solution to problem (7.2) if G has negative eigenvalues. An easy normalization may work as follows. Since G is symmetric, it can be written in terms of its spectral decomposition

$$
G = V \Lambda V^T,
$$

where $\Lambda = \text{diag}(\lambda_1, \lambda_2, \ldots, \lambda_n)$ is the diagonal matrix of eigenvalues of G, and $V = [v_1, \ldots, v_n]$ is the matrix of corresponding eigenvectors. Then $K' = V \hat{\Lambda} V^T$, where $\hat{\Lambda} = \text{diag}(\max(\lambda_1, 0), \ldots, \max(\lambda_n, 0))$, is the solution to problem (7.2). Let

$$
k_{ij} = \frac{k'_{ij}}{\sqrt{k'_{ii} k'_{jj}}},
$$

and we have $k_{ii} = 1$. A potential problem with this method is that k_{ii} may be close to 0 for some i in practice.

In summary, we formulate the nearest positive semidefinite matrix (NPSDM) problem as follows:

$$
\begin{aligned}
&\min \|G - K\|^2 \\
&\text{subject to } K \succeq 0 \\
&\qquad\qquad K_{ii} = 1.
\end{aligned}
\tag{7.3}
$$

It is widely known that this problem can be reformulated as a semidefinite programming problem by introducing new variables:

min t

subject to

$$
\begin{pmatrix} K & 0 & 0 \\ 0 & I_{n^2} & vec(G) - vec(K) \\ 0 & (vec(G) - vec(K))^T & t \end{pmatrix} \succeq 0,
$$

$$
K_{ii} = 1
$$

where the vec operator stacks the columns of a matrix into one long vector. As counted in [Hig02], in total there are $n^4/2 + 3n^2/2 + n + 1$ constraints in this SDP formulation, where n is the number of training documents in our case. [Hig02] further commented on this reformulation that "unfortunately, this number of constraints makes it impractical to apply a general semidefinite programming solver—merely specifying the constraints (taking full advantage of their sparsity) requires prohibitive amount of memory." Numerical experiments with the SeDuMi package [Stu99] against problems of $n \geq 50$ confirm this conclusion.

With a semidefinite matrix K, we can apply the kernel method and MDS on XML document representation.

7.5 Newton-Type Method

In this section, we will show how a Newton-type method in [QS06] can be applied to the problem (7.3). Actually, problem (7.3) is another version of the low-rank approximation problem which is nonconvex and extremely hard to solve.

Given a symmetric matrix G, computing its nearest correlation matrix is recently studied by using the projection method in [Hig02]. The projection method is known to be linearly convergent. When the underlying problem is big, the projection method is quite slow, requiring many iterations to terminate. To introduce the semismooth techniques in our Newton's method, we rewrite problem (7.3) as

$$
\begin{aligned}
\min \; & \| G - K \|^2 \\
\text{s.t. } & K_{ii} = 1, \;\; i = 1, \dots, n \\
& K \in \mathcal{S}_+,
\end{aligned}
\tag{7.4}
$$

where $\mathcal{S}_+$ is the cone of positive semidefinite matrices in $\mathcal{S}$, the set of $n \times n$ symmetric matrices.

The Newton method for (7.4) is based on its dual problem rather than on itself. According to Rockafellar [Roc74], the dual problem is the following unconstrained and differentiable problem:

$$
\min_{z \in \Re^n} \theta(z) := \frac{1}{2} \| (G + \mathcal{A}^* z)_+ \|^2 - b^T z,
\tag{7.5}
$$

where $\mathcal{A} : \mathcal{S} \mapsto \Re^n$ is the diagonal operator defined by $\mathcal{A}(K) = \mathrm{diag}(K)$ and $\mathcal{A}^*$ is its adjoint operator, $b = e$ the vector of all ones, and for a matrix $K \in \mathcal{S}$, K_+ denotes its orthogonal projection to $\mathcal{S}_+$. We note that $\theta(\cdot)$ is only once continuously differentiable. Hence, quasi-Newton methods are suitable to find a solution of (7.5), see [Mal05, BX05]. However, the convergence rate of these methods appears to be only linear at best. A fast convergent method is more desirable in order to save computation given that the variables are of matrices with positive semidefinite constraints to satisfy.

The optimality condition of the dual problem (7.5) is

$$
\mathcal{A} (G + \mathcal{A}^* z)_+ = b, \quad z \in \Re^n.
\tag{7.6}
$$

Once a solution z^* of (7.6) is found, we can construct the optimal solution of (7.4) by

$$K^* = (G + \mathcal{A}^* z^*)_+ . \tag{7.7}$$

A very important question on K^* is when it is of full rank (i.e., $\text{rank}(K^*) = n$). Equivalently, one may ask when K^* is rank-deficient (i.e., $\text{rank}(K^*) < n$). An interesting interpretation of the rank of K^* is that the number $\text{rank}(K^*)$ represents the defining factors from all n factors. These defining factors usually characterize the fundamental structure of the space that consists of all n factors. It remains to see how the low rank approximation K^* can be interpreted for the problem (7.3).

To present a result characterizing the rank of K^*, we need some notations. Let

$$\mathcal{U} := \{Y \in \mathcal{S} : Y_{ii} = 1, \ i = 1, \ldots, n\}.$$

Then K is a positive semidefinite matrix if and only if $K \in \mathcal{S}_+ \cap \mathcal{U}$. For $K, Y \in \mathcal{S}$, the inner product of K and Y is defined by

$$\langle K, Y \rangle = \text{Trace}(KY).$$

The normal cone of the convex sets $\mathcal{S}_+$ and $\mathcal{U}$ at a point $K \in S$ are given as follows [Hig02, Lemmas 2.1 and 2.2]:

$$\partial \mathcal{S}_+(K) = \{Y \in \mathcal{S} : \langle Y, K \rangle = 0, \ Y \preceq 0\}$$
$$\partial \mathcal{U}(K) = \{\text{diag}(\eta_i) : \ \eta_i \text{ arbitrary}\}.$$

The following result characterizes when K^* is of full rank.

Proposition 1. *Let $G \in \mathcal{S}$ be given and K^* denotes its nearest positive semidefinite matrix. Let $\mathcal{C}_{++}$ denote the set of all positive definite matrices and $\mathcal{D}$ denote the set of all diagonal matrices. Then $\text{rank}(K^*) = n$ if and only if*

$$G \in \mathcal{D} + \mathcal{C}_{++}.$$

Proof. It follows from [Hig02, Theorem 2.4] that a positive semidefinite matrix K^* solves (7.4) if and only if

$$K^* = G + (VDV^T + \text{diag}(\tau_i)),$$

where $V \in \Re^{n \times n}$ has orthonormal columns spanning the null space of K^*, $D = \text{diag}(d_i) \geq 0$, and the τ_i are arbitrary.

Suppose now that $\text{rank}(K^*) = n$ (i.e., $K^* \in \mathcal{C}_{++}$), then $V = 0$, which yields

$$K^* = G + \text{diag}(\tau_i).$$

Equivalently,

$$G = K^* - \text{diag}(\tau_i) \in \mathcal{C}_{++} + \mathcal{D}.$$

This completes the necessary part of the proof.

Now suppose we have $G \in \mathcal{C}_{++} + \mathcal{D}$. Then there exists a diagonal matrix D and a positive definite matrix $Y \in \mathcal{C}_{++}$ such that

$$G = D + Y. \tag{7.8}$$

It follows from the characterization of K^* [Hig02, Eq. 2.4] that

$$G - K^* \in \partial S_+(K^*) + \partial \mathcal{U}(K^*).$$

Hence there exists a matrix Z satisfying $Z \succeq 0$ and $\langle Z, K^* \rangle = 0$, and a diagonal matrix $D' \in \partial U(K^*)$, such that

$$G - K^* = -Z + D'.$$

Noting (7.8), we have

$$D - D' = K^* - Y - Z,$$

which gives

$$K^* = D - D' + Y + Z.$$

Because both K^* and Y are positive semidefinite matrices, their diagonal elements are all equal to 1. Hence,

$$D - D' = -\mathrm{diag}(Z),$$

where $\mathrm{diag}(Z)$ is the diagonal matrix whose diagonal elements are that of Z.

By using the relation that $\langle K^*, Z \rangle = 0$ and relations above, we have

$$\begin{aligned}
0 &= \langle Z, K^* \rangle \\
&= \langle Z, D - D' \rangle + \langle Z, Y \rangle + \|Z\|^2 \\
&= -\langle Z, \mathrm{diag}(Z) \rangle + \langle Z, Y \rangle + \|Z\|^2 \\
&\geq \langle Z, Y \rangle.
\end{aligned}$$

The inequality uses the fact that $\|Z\|^2 - \langle Z, \mathrm{diag}(Z) \rangle \geq 0$. Since Y is positive definite and Z is positive semidefinite, the fact that $\langle Z, Y \rangle \leq 0$ means that $Z = 0$, which in turn implies $D - D' = 0$. Hence

$$K^* = D - D' + Y + Z = Y$$

and $\mathrm{rank}(K^*) = \mathrm{rank}(Y) = n$, completing the sufficient part of the proof.

Proposition 1 provides an interesting justification on a heuristic method to find a positive semidefinite matrix from a given matrix G. Specifically, we replace each of the diagonal elements of G with 1, denoted by $\tilde{G}$, and conduct a spectral decomposition of $\tilde{G}$. If $\tilde{G}$ is a positive semidefinite matrix, then it is all done. Otherwise we will project $\tilde{G}$ to the semidefinite matrix cone S_+ and get a new matrix, denoted by $\tilde{G}_+$. Replace all the diagonal elements of $\tilde{G}_+$ by *one* and test it if it is a positive semidefinite matrix. The procedure carries on until a positive semidefinite matrix is found.

The method tries to solve the optimality Eq. (7.6). For notation convenience, let

$$F(z) := \mathcal{A}(G + \mathcal{A}^* z)_+.$$

Then Eq. (7.6) becomes

$$F(z) = b. \tag{7.9}$$

Since F is not differentiable due to the projection operator $(\cdot)_+$, the classic Newton method is not suitable. However, we may use the generalized Newton method to solve it.

$$z^{k+1} = z^k - V_k^{-1}(F(z^k) - b), \tag{7.10}$$
$$V_k \in \partial F(z^k), k = 0, 1, 2, \ldots$$

where $\partial F(z)$ denotes the generalized Jacobian of F, see [Cla83].

As analyzed in [QS06], the generalized Newton method has the following properties. First, the function F is strongly semismooth everywhere, a crucial property for the method to be convergent. Due to this reason, the method (7.10) is often referred to as the semismooth Newton method. Second, every element in $\partial F(z^*)$ is nonsingular, where z^* is a solution of (7.9). This nonsingularity property justifies why we can calculate the inverse of V_k when z^k is near z^*. The consequence of these properties is Theorem 2.

Theorem 2. *[QS06] The semismooth Newton method (7.10) is locally quadratically convergent.*

7.6 Experimental Studies

In this section, we examine the effectiveness of the kernel-based approaches on the classification of structured data. The goal is to evaluate the classification accuracy of the kernel-based approaches and compare the nearest semidefinite matrix method with the method using the original similarity matrix without correction.

The dataset used in our experiments is obtained from the MovieDB corpus, which was created by using the IMDB [DG] database. The dataset contains 9643 XML documents. MovieDB is designed for both the structure only, and structure and content tasks. Since our algorithm exploits only structural information, all labels of the documents come from 11 possible structure categories, which correspond to transformations of the original data structure. There are four classification tasks, created with different levels of noise on the structure. Training and test sets are generated by randomly selecting samples from six of 11 categories. At two different noise levels, we generated five training sets and five test sets, using 5, 10, 20, 30, and 40 samples, respectively, in each selected categories.

To construct kernel matrices over the five sets of samples, we first infer their schema and then compute similarities among these documents according to Eq. (7.1). Similar to the Xtract scheme, we extract candidate regular expressions based on the frequencies of the repeat patterns appearing in the input sequences. The candidate regular expressions are ranked by using the MLD principle. To make the Xtract scheme usable on large-size document sets, we modify it as follows:

1. The frequencies of the child sequences are evaluated with some victims not covered by the inferred regular expressions. The victims are those sequences that appears very infrequently. We find that this modification can reduce the negative effects of noises in classification.
2. The aim of the modification is to minimize $L = \lambda L_g + L_d$, where λ is the weight to balance the preciseness and generality of the schema. We develop a practical scheme to manipulate λ.
3. We use the cost of nondeterministic finite automata (NFA) simulation rather than enumerating multiple sequence partitions to compute the minimum cost of encoding so that the computation complexity is reduced.

After the schema have been extracted, we construct the similarity matrices by computing the similarities among samples, resulting in five matrices (30×30, 60×60, 120×120, 180×180, 240×240) for each noise level.

The implementation of the nearest semidefinite matrix method is supported by the LibSVM software package [CL01] and codes for the Newton-type method [QS06]. All experimental studies were carried out on a PC with Pentium IV 3.4GHz CPU, 2 GB of memory. The results obtained for two different noise levels are given in Table 7.1 and Table 7.2, respectively.

As we can see in Table 7.1, the accuracy of classification by MDS and the kernel method has been improved over almost each dataset. We note that the results depend on the number of samples available and how to select them. As shown in the table, the correctness increases generally when the number of samples increases. However, the accuracy for dataset 120×120 drops a little because more noise data are chosen in this group. The datasets in Table 7.2 are presented with more noise than those in Table 7.1. Hence, the accuracy of each dataset in Table 7.2 is much lower than that of Table 7.1. The results in both tables suggest that the nearest semidefinite matrix method can improve the performance of the kernel-based SVM classifier.

Table 7.1. Classification accuracy (%) for datasets at noise level 1

Methods	30	60	120	180	240
Similarity matrix	87.23	92.67	91.33	96.67	97.33
MDS	89.33	94.0	96.67	98.33	98.33
Kernel method	94.0	96.67	94.0	97.33	100

Table 7.2. Classification accuracy (%) for datasets at noise level 3

Methods	30	60	120	180	240
Similarity matrix	63.33	68.33	71.67	70.0	73.33
MDS	66.67	73.33	73.33	71.67	73.33
NPSD matrix	71.67	73.33	76.67	71.67	80.33

7.7 Conclusion and Discussion

In this chapter, we have studied two data representation methods for classifying XML documents based on their structures. Based on the fact that many XML documents are without schema, we first infer the document schema given a set of documents and then estimate the similarities among the documents. Since the estimated similarity matrix is usually not a distance matrix, it is not a well-defined kernel matrix. We formulate the problem as a nearest semidefinite matrix problem and present a novel Newton-type method, named NPSDM, to determine the similarity matrix for the set of XML documents. The NPSDM is quadratically convergent and usually takes no more than 10 iterations to converge. Using the Newton-type method, it has been reported recently [QS06] that problems with n up to 2000 have been successfully solved. To make use of the Newton method, we have to address how to select an element V_k in $\partial F(z^k)$. One of the choices can be found in [QS06]. Another issue is to globalize the method in Eq. (7.10), which in its current form is only locally convergent as indicated in the above theorem. One globalized version can also be found in [QS06, Algorithm 5.1].

Acknowledgment

This work was supported in part by the Kentucky Experimental Program to Stimulate Competitive Research (EPSCoR) under grant No. 4800000786.

References

[Bur98]   C.J.C. Burges. A tutorial on support vector machines for pattern recognition. *Data Min. Knowl. Discov.*, 2(2):121–167, 1998.

[Bur05]   C.J.C. Burges. Geometric methods for feature extraction and dimensional reduction — a guided tour. In *The Data Mining and Knowledge Discovery Handbook*, pages 59–92. Springer, New York, 2005.

[BX05]   S. Boyd and L. Xiao. Least-squares covariance matrix adjustment. *SIAM Journal on Matrix Analysis and Applications*, 27(2):532–546, 2005. Available from World Wide Web: http://link.aip.org/link/?SML/27/532/1.

[CC01]   T.F. Cox and M.A.A. Cox. *Multidimensional Scaling*. Monographs on Statistics and Applied Probability. Chapman & Hall/CRC, Boca Raton, 2nd edition, 2001.

[Che01]   W. Chen. New algorithm for ordered tree-to-tree correction problem. *J. Algorithms*, 40(2):135–158, 2001.

[CL01]   C.-C. Chang and C.-J. Lin. *LIBSVM: a library for support vector machines*, 2001. Software available at http://www.csie.ntu.edu.tw/~cjlin/libsvm.

[Cla83]   F.H. Clark. *Optimization and Nonsmooth Analysis*. John Wiley and Sons, New York, 1983.

[CX05]   E.R. Canfield and G. Xing. Approximate xml document matching. In *SAC '05: Proceedings of the 2005 ACM Symposium on Applied Computing*, pages 787–788. ACM Press, New York, 2005.

[DG]   L. Denoyer and P. Gallinari. *XML Document Mining Challenge.* Database available at http://xmlmining.lip6.fr/.

[DG04]   L. Denoyer and P. Gallinari. Bayesian network model for semistructured document classification. *Inf. Process. Manage.*, 40(5):807–827, 2004.

[GGR⁺00] M. Garofalakis, A. Gionis, R. Rastogi, S. Seshadri, and K. Shim. Xtract: a system for extracting document type descriptors from xml documents. In *SIGMOD '00: Proceedings of the 2000 ACM SIGMOD International Conference on Management of Data*, pages 165–176. ACM Press, New York, 2000.

[GJK⁺06] S. Guha, H.V. Jagadish, N. Koudas, D. Srivastava, and T. Yu. Integrating xml data sources using approximate joins. *ACM Trans. Database Syst.*, 31(1):161–207, 2006.

[Hig02]   N.J. Higham. Computing the nearest correlation matrix — a problem from finance. *IMA Journal of Numerical Analysis*, 22(3):329–343, 2002.

[Joa98]   T. Joachims. Text categorization with suport vector machines: learning with many relevant features. In Claire Nédellec and Céline Rouveirol, editors, *Proceedings of ECML-98, 10th European Conference on Machine Learning*, volume 1398 of *Lecture Notes in Computer Science*, pages 137–142. Springer, New York, 1998. Available from World Wide Web: citeseer.ist.psu.edu/joachims97text.html.

[LCB⁺04] G.R.G. Lanckriet, N. Cristianini, P. Bartlett, L. El Ghaoui, and M.I. Jordan. Learning the kernel matrix with semidefinite programming. *J. Mach. Learn. Res.*, 5:27–72, 2004.

[LCMY04] W. Lian, D.W. Cheung, N. Mamoulis, and S.-M. Yiu. An efficient and scalable algorithm for clustering xml documents by structure. *IEEE Transactions on Knowledge and Data Engineering*, 16(1):82–96, 2004.

[Mal05]   J. Malick. A dual approach to semidefinite least-squares problems. *SIAM J. Matrix Anal. Appl.*, 26(1):272–284, 2005.

[Mur00]   M. Murata. Hedge automata: a formal model for XML schemata. Web page, 2000. Available from World Wide Web: citeseer.ist.psu.edu/article/murata99hedge.html.

[NJ02]   A. Nierman and H.V. Jagadish. Evaluating structural similarity in xml documents. In *WebDB*, pages 61–66, 2002.

[QS06]   H. Qi and D. Sun. A quadratically convergent newton method for computing the nearest correlation matrix. *SIAM J. Matrix Anal. Appl.*, 28(2):360–385, 2006.

[Roc74]   R.T. Rockafellar. *Conjugate duality and optimization.* Society for Industrial and Applied Mathematics, Philadelphia, 1974.

[Seb02]   F. Sebastiani. Machine learning in automated text categorization. *ACM Comput. Surv.*, 34(1):1–47, 2002.

[Stu99]   J.F. Sturm. Using SeDuMi 1.02, a MATLAB toolbox for optimization over symmetric cones. *Optimization Methods and Software*, 11-12:625–653, 1999. Available from World Wide Web: citeseer.ist.psu.edu/sturm99using.html. Special issue on Interior Point Methods (CD supplement with software).

[STV04]   B. Schölkopf, K. Tsuda, and J.P. Vert. *Kernel Methods in Computational Biology.* MIT Press, Cambridge, MA, 2004.

[SZ97]   D. Shasha and K. Zhang. Approximate tree pattern matching. In *Pattern Matching Algorithms*, pages 341–371. Oxford University Press, New York, 1997. Available from World Wide Web: citeseer.ist.psu.edu/shasha95approximate.html.

[SZC+04] J.-T. Sun, B.-Y. Zhang, Z. Chen, Y.-C. Lu, C.-Y. Shi, and W.-Y. Ma. Ge-cko: A method to optimize composite kernels for web page classification. In *WI '04: Proceedings of the 2004 IEEE/WIC/ACM International Conference on Web Intelligence*, pages 299–305, 2004.

[ZA03]   M.J. Zaki and C.C. Aggarwal. Xrules: an effective structural classifier for xml data. In *KDD '03: Proceedings of the Ninth ACM SIGKDD International Conference on Knowledge Discovery and Data Mining*, pages 316–325. ACM Press, New York, 2003.

[ZI02]   T. Zhang and V.S. Iyengar. Recommender systems using linear classifiers. *J. Mach. Learn. Res.*, 2:313–334, 2002.

Part III

Email Surveillance and Filtering

8

Discussion Tracking in Enron Email Using PARAFAC

Brett W. Bader, Michael W. Berry, and Murray Browne

Overview

In this chapter, we apply a nonnegative tensor factorization algorithm to extract and detect meaningful discussions from electronic mail messages for a period of one year. For the publicly released Enron electronic mail collection, we encode a sparse term-author-month array for subsequent three-way factorization using the PARAllel FACtors (or PARAFAC) three-way decomposition first proposed by Harshman. Using nonnegative tensors, we preserve natural data nonnegativity and avoid subtractive basis vector and encoding interactions present in techniques such as principal component analysis. Results in thread detection and interpretation are discussed in the context of published Enron business practices and activities, and benchmarks addressing the computational complexity of our approach are provided. The resulting tensor factorizations can be used to produce Gantt-like charts that can be used to assess the duration, order, and dependencies of focused discussions against the progression of time.

8.1 Introduction

When Enron closed its doors on December 2, 2001, and filed for Chapter 11 bankruptcy, it began a turn of events that released an unprecedented amount of information (over 1.5 million electronic mail messages, phone tapes, internal documents) into the public domain. This information was the cornerstone of the Federal Energy Regulatory Commission's (FERC) investigation into the global energy corporation. The original set of emails was posted on FERC's Web site [Gri03], but it suffered document integrity problems, and attempts were made to improve the quality of the data and remove sensitive and irrelevant private information. Dr. William Cohen of Carnegie-Mellon University oversaw the distribution of this improved corpus— known as the Enron email sets. The latest version of the Enron email sets[1] (dated

---

[1] http://www-2.cs.cmu.edu/~enron

March 2, 2004) contains $517,431$ email messages of 150 Enron employees covering a period from December 1979 through February 2004, with the majority of the messages spanning the three years of 1999, 2000, and 2001.

For the most part, the emails reflect the day-to-day activities of America's seventh largest company, but certain distinct topics of discussion are linked to Enron. One involved Enron's development of the Dabhol Power Company (DPC) in the Indian state of Maharashtra, an endeavor awash in years of logistical and political problems. Another topic was the deregulation of the California energy market, which led to rolling blackouts during the summer of 2000—a situation that Enron (and other energy companies) took advantage of financially. Eventually a combination of greed, overspeculation, and deceptive accounting practices snowballed into an abrupt collapse in the fourth quarter of 2001, which again is reflected in the emails. The Enron email sets provide a unique opportunity not only to study the mechanics of a sizable email network, but it also offers a glimpse of the machinations of how huge global corporations operate on a day-to-day basis.

In this research, we seek to extract meaningful threads of discussion from a subset of the Enron email set. The idea underlying our thread extraction is as follows. Suppose we have a collection of q emails from n authors over a period of p months (or similar unit of time). In aggregate, there are a collection of m terms parsed from the q emails. From these data, we create an $m \times n \times p$ term-author-month array[2] $\mathcal{X}$. We then decompose $\mathcal{X}$ using PARAFAC or a nonnegative tensor factorization to track discussions over time.

In the next section we provide background information on tensor decompositions and related work. Section 8.3 provides a formal discussion of the notations used to define these decompositions and algorithms that are given in Section 8.4. Details of the specific Enron subset used in this study are provided in Section 8.5, and our observations and results in applying PARAFAC to this subset of emails follow in Section 8.6. Finally, a brief summary of future work in the use of nonnegative tensor factorization for discussion tracking is given in Section 8.7.

8.2 Related Work

Tensor decompositions date back 40 years [Tuc66, Har70, CC70], and they have been used extensively in a variety of domains, from chemometrics [SBG04] to signal processing [SGB00]. PARAFAC is a three-way decomposition that was proposed by Harshman [Har70] using the name PARAllel FACtors or PARAFAC. At the same time, Carroll and Chang [CC70] published the same mathematical model, which they call canonical decomposition (CANDECOMP).

The use of multidimensional models is relatively new in the context of text analysis. Acar et al. [AÇKY05] use various tensor decompositions of (user × key word × time) data to separate different streams of conversation in chatroom data. Several

---

[2] Note that the array $\mathcal{X}$ is generally sparse due to the word distribution used by each author over time.

Web search applications involving tensors relied on query terms or anchor text to provide a third dimension. Sun et al. [SZL$^+$05] apply a three-way Tucker decomposition [Tuc66] to the analysis of (user $\times$ query term $\times$ Web page) data in order to personalize Web searches. Kolda et al. [KBK05] and Kolda and Bader [KB06] use PARAFAC on a (Web page $\times$ Web page $\times$ anchor text) sparse, three-way tensor representing the Web graph with anchor-text-labeled edges to get hub/authority rankings of pages related to an identified topic.

8.3 Notation

Multidimensional arrays and tensors are denoted by boldface Euler script letters, for example, $\boldsymbol{\mathfrak{X}}$. Element (i, j, k) of a third-order tensor $\boldsymbol{\mathfrak{X}}$ is denoted by x_{ijk}.

The symbol $\circ$ denotes the tensor outer product,

$$A_1 \circ B_1 = \begin{pmatrix} A_{11}B_{11} & \cdots & A_{11}B_{m1} \\ \vdots & \ddots & \vdots \\ A_{m1}B_{11} & \cdots & A_{m1}B_{m1} \end{pmatrix}.$$

The symbol $*$ denotes the Hadamard (i.e., elementwise) matrix product,

$$A * B = \begin{pmatrix} A_{11}B_{11} & \cdots & A_{1n}B_{1n} \\ \vdots & \ddots & \vdots \\ A_{m1}B_{m1} & \cdots & A_{mn}B_{mn} \end{pmatrix}.$$

The symbol $\otimes$ denotes the Kronecker product,

$$A \otimes B = \begin{pmatrix} A_{11}B & \cdots & A_{1n}B \\ \vdots & \ddots & \vdots \\ A_{m1}B & \cdots & A_{mn}B \end{pmatrix}.$$

The symbol $\odot$ denotes the Khatri-Rao product (columnwise Kronecker) [SBG04],

$$A \odot B = \begin{pmatrix} A_1 \otimes B_1 & \cdots & A_n \otimes B_n \end{pmatrix}.$$

The concept of *matricizing* or *unfolding* is simply a rearrangement of the entries of $\boldsymbol{\mathfrak{X}}$ into a matrix. Although different notations exist, we are following the notation used in [SBG04]. For a three-dimensional array $\boldsymbol{\mathfrak{X}}$ of size $m \times n \times p$, the notation $X^{(m \times np)}$ represents a matrix of size $m \times np$ in which the n-index runs the fastest over the columns and p the slowest. Other permutations, such as $X^{(p \times nm)}$, are possible by changing the row index and the fast/slow column indices.

The norm of a tensor, $\| \boldsymbol{\mathfrak{X}} \|$, is the same as the Frobenius norm of the matricized array, that is, the square root of the sum of squares of all its elements.

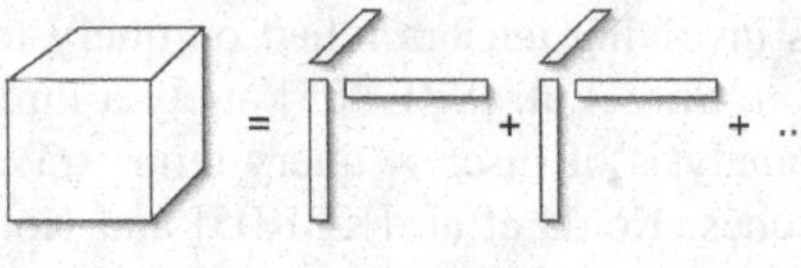

Fig. 8.1. PARAFAC provides a three-way decomposition with some similarity to the singular value decomposition.

8.4 Tensor Decompositions and Algorithms

Suppose we are given a tensor $\mathcal{X}$ of size $m \times n \times p$ and a desired approximation rank r. The goal is to decompose $\mathcal{X}$ as a sum of vector outer products as shown in Figure 8.1. It is convenient to group all r vectors together in factor matrices A, B, C, each having r columns. The following mathematical expressions of this model use different notations but are equivalent:

$$x_{ijk} \approx \sum_{l=1}^{r} A_{il} B_{jl} C_{kl},$$

$$\mathcal{X} \approx \sum_{l=1}^{r} A_l \circ B_l \circ C_l, \tag{8.1}$$

$$X^{(m \times np)} \approx A(C \odot B)^T.$$

PARAFAC may apply to general N-way data, but because our application pertains only to three-way data, we are considering only the specific three-way problem at this time.

Without loss of generality, we typically normalize all columns of the factor matrices to have unit length and store the accumulated weight (i.e., like a singular value) in a vector λ:

$$\mathcal{X} \approx \sum_{l=1}^{r} \lambda_l (A_l \circ B_l \circ C_l)$$

Moreover, we typically reorder the final solution so that $\lambda_1 \geq \lambda_2 \geq \cdots \geq \lambda_r$. In the following subsections, we describe general algorithms for the model without λ because this normalization can be performed in a postprocessing step.

Our goal is to find the best-fitting matrices A, B, and C in the minimization problem:

$$\min_{A,B,C} \left\| \mathcal{X} - \sum_{l=1}^{r} A_l \circ B_l \circ C_l \right\|^2. \tag{8.2}$$

It is important to note that the factor matrices are not required to be orthogonal. Under mild conditions, PARAFAC provides a unique solution that is invariant to factor rotation [Har70]. Hence, the factors are plausibly a valid description of the data with greater reason to believe that they have more explanatory meaning than a "nice" rotated two-way solution.

Given a value $r > 0$ (loosely corresponding to the number of distinct topics or conversations in our data), the tensor decomposition algorithms find matrices $A \in \mathbb{R}^{m \times r}$, $B \in \mathbb{R}^{n \times r}$, and $C \in \mathbb{R}^{p \times r}$, to yield Equation (8.1). Each triad $\{A_j, B_j, C_j\}$, for $j = 1, \ldots, r$, defines scores for a set of terms, authors, and time for a particular conversation in our email collection; the value λ_r after normalization defines the weight of the conversation. (Without loss of generality, we assume the columns of our matrices are normalized to have unit length.) The scales in C indicate the activity of each conversation topic over time.

8.4.1 PARAFAC-ALS

A common approach to solving Equation (8.2) is an alternating least squares (ALS) algorithm [Har70, FBH03, TB05], due to its simplicity and ability to handle constraints. At each inner iteration, we compute an entire factor matrix while holding all the others fixed.

Starting with random initializations for A, B, and C, we update these quantities in an alternating fashion using the method of normal equations. The minimization problem involving A in Equation (8.2) can be rewritten in *matrix* form as a least squares problem [FBH03]:

$$\min_A \left\| X^{(m \times np)} - AZ \right\|^2, \tag{8.3}$$

where $Z = (C \odot B)^T$.

The least squares solution for Equation (8.3) involves the pseudo-inverse of Z:

$$A = X^{(m \times np)} Z^\dagger.$$

Conveniently, the pseudo-inverse of Z may be computed in a special way that avoids computing $Z^T Z$ with an explict Z [SBG04], so the solution to Equation (8.3) is given by

$$A = X^{(m \times np)}(C \odot B)(B^T B * C^T C)^{-1}.$$

Furthermore, the product $X^{(m \times np)}(C \odot B)$ may be computed efficiently if $\mathfrak{X}$ is sparse [KB06] by not forming the Khatri-Rao product $C \odot B$. Thus, computing A essentially reduces to several matrix inner products, sparse tensor-matrix multiplication of B and C into $\mathfrak{X}$, and inverting an $R \times R$ matrix.

Analogous least-squares steps may be used to update B and C.

8.4.2 Nonnegative Tensor Factorization

We also considered a PARAFAC model with nonnegativity constraints on the factor matrices. Because we are dealing with nonnegative data in $\mathfrak{X}$, it often helps to examine decompositions that retain the nonnegative characteristics of the original data. Modifications to the ALS algorithm are needed, and we use the multiplicative update introduced in [LS99] and adapted for tensor decompositions by Mørup

[Mør05, MHA06]. We also incorporate the addition of ϵ for stability as was done in [BB05a]. Overall, the approach is similar to PARAFAC-ALS except that the factor matrices are updated differently.

First, we note that residual norm of the various formulations of the PARAFAC model are equal:

$$\|X^{(m \times np)} - A(C \odot B)^T\|_F =$$
$$\|X^{(n \times mp)} - B(C \odot A)^T\|_F =$$
$$\|X^{(p \times mn)} - C(B \odot A)^T\|_F.$$

Each of these matrix systems is treated as a nonnegative matrix factorization (NMF) problem and solved in an alternating fashion. That is, we solve for A using the multiplicative update rule holding B and C fixed, and so on:

$$A_{i\rho} \leftarrow A_{i\rho} \frac{(X^{(m \times np)} Z)_{i\rho}}{(AZ^T Z)_{i\rho} + \epsilon}, \quad Z = (C \odot B)$$

$$B_{j\rho} \leftarrow B_{j\rho} \frac{(X^{(n \times mp)} Z)_{j\rho}}{(BZ^T Z)_{j\rho} + \epsilon}, \quad Z = (C \odot A)$$

$$C_{k\rho} \leftarrow C_{k\rho} \frac{(X^{(p \times mn)} Z)_{k\rho}}{(CZ^T Z)_{k\rho} + \epsilon}, \quad Z = (B \odot A)$$

Here ϵ is a small number like 10^{-9} that adds stability to the calculation and guards against introducing a negative number from numerical underflow.

As was mentioned previously, $\mathcal{X}$ is sparse, which facilitates a simpler computation in the procedure above. Each matricized version of $\mathcal{X}$ (which has the same nonzeros but reshaped) is a sparse matrix. The matrix Z from each step should not be formed explicitly because it would be a large, dense matrix. Instead, the product of a matricized $\mathcal{X}$ with Z should be computed specially, exploiting the inherent Kronecker product structure in Z so that only the required elements in Z need to be computed and multiplied with the nonzero elements of $\mathcal{X}$. See [KB06] for details.

8.5 Enron Subset

For a relevant application, we consider the email corpus of the Enron corporation that was made public during the federal investigation. The whole collection is available online [Coh] and contains 517,431 emails stored in the mail directories of 150 users. We use a smaller graph of the Enron email corpus prepared by Priebe et al. [PCMP06] that consists of messages among 184 Enron email addresses plus 13 more that have been identified in [BB05a] as interesting. We considered messages only in 2001, which resulted in a total of 53,733 messages over 12 months (messages were sent on a total of 357 days).

An obvious difficulty in dealing with the Enron corpus is the lack of information regarding the former employees. Without access to a corporate directory or organizational chart of Enron at the time of these emails, it is difficult to ascertain the validity of our results and assess the performance of the PARAFAC model. Other researchers using the Enron corpus have had this same problem, and information on the participants has been collected slowly and made available.

The Priebe dataset [PCMP06] provided partial information on the 184 employees of the small Enron network, which appears to be based largely on information collected by Shetty and Adibi [SA05]. It provides most employees' position and business unit. To facilitate a better analysis of the PARAFAC results, we collected extra information on the participants from the email messages themselves and found some relevant information posted on the FERC Web site [Fed]. To help assess our results, we searched for corroborating information of the preexisting data or for new identification information, such as title, business unit, or manager. Table 8.1 lists 11 of the most notable authors (and their titles) whose emails were tracked in this study.

Table 8.1. Eleven of the 197 email authors represented in the term-author-time array $\mathcal{X}$

Name	Email account (@enron.com)	Title
Richard Sanders	b..sanders	VP Enron Wholesale Services
Greg Whalley	greg.whalley	President
Jeff Dasovich	jeff.dasovich	Employee Government Relationship Executive
Jeffery Skilling	jeff.skilling	CEO
Steven Kean	j..kean	VP and Chief of Staff
John Lavorato	john.lavorato	CEO Enron America
Kenneth Lay	kenneth.lay	CEO
Louise Kitchen	louise.kitchen	President Enron Online
Mark Haedicke	mark.haedicke	Managing Director Legal Department
Richard Shapiro	richard.shapiro	VP Regulatory Affairs
Vince Kaminski	vince.kaminski	Manager Risk Management Head, Enron Energy Services

Of the 197 authors whose emails were tracked (in the year 2001), there were a few cases of aliasing. That is, different email accounts of the form

employee_id@enron.com

were used by the same employee. A few sample aliases from the 11 notable authors in Table 8.1 are David Delaney (david.delainey, w..delainey) and Vince Kaminski (j.kaminski, j..kaminski, vince.kaminski).

8.5.1 Term Weighting

We considered two datasets: monthly and daily time periods. The monthly data correspond to a sparse adjacency array $\mathcal{X}$ of size $69157 \times 197 \times 12$ with 1,042,202 nonzeros, and the daily data correspond to a sparse adjacency array $\mathcal{Y}$ of size $69157 \times 197 \times 357$ with 1,770,233 nonzeros. The 69,157 terms were parsed from the 53,733 messages using a master dictionary of 121,393 terms created by the General

Text Parser (GTP) software environment (in C++) maintained at the University of Tennessee [GWB03]. This larger set of terms was previously obtained when GTP was used to parse 289,695 of the 517,431 emails defining the Cohen distribution at Carnegie-Mellon (see Section 8.1). To be accepted into the dictionary, a term had to occur in more than one email and more than 10 times among the 289,695 emails.

Unique to previous parsings of Enron subsets by GTP (see [BBL+07, SBPP06, BB05b]), a much larger *stoplist* of unimportant words was used to filter out the content-rich 69,157 terms for the $\mathcal{X}$ and $\mathcal{Y}$ arrays. This stoplist of 47,154 words was human-generated by careful screening of the master (GTP-generated) dictionary for words with no specific reference to an Enron-related person or activity.

We scaled the nonzero entries of $\mathcal{X}$ and $\mathcal{Y}$ according to a weighted frequency:

$$x_{ijk} = l_{ijk} g_i a_j,$$
$$y_{ijk} = l_{ijk} g_i a_j,$$

where l_{ijk} is the local weight for term i written by author j in month/day k, g_i is the global weight for term i, and a_j is an author normalization factor.

Let f_{ijk} be the number of times term i is written by author j in month/day k, and define $h_{ij} = \frac{\sum_k f_{ijk}}{\sum_{jk} f_{ijk}}$. The specific components of each nonzero are listed below:

$$\text{Log local weight} \qquad l_{ijk} = \log(1 + f_{ijk})$$

$$\text{Entropy global weight } g_i = 1 + \sum_{j=1}^{n} \frac{h_{ij} \log h_{ij}}{\log n}$$

$$\text{Author normalization } a_j = \frac{1}{\sqrt{\sum_{i,k} (l_{ijk} g_i)}}$$

These weights are adapted from the well-known log-entropy weighting scheme [BB05c] used on term-by-document matrices. The log local weight scales the raw term frequencies to diminish the importance of high-frequency terms. The entropy global weight attempts to emulate an entropy calculation of the terms over all messages in the collection to help discriminate important terms from common terms. The author normalization helps to correct imbalances in the number of messages sent from each author. Without some type of normalization, discussions involving prolific authors would tend to dominate the results.

8.6 Observations and Results

In this section we summarize our findings of applying PARAFAC and nonnegative tensor factorization (NNTF) on the Enron email collection. Our algorithms were written in MATLAB, using sparse extensions of the Tensor Toolbox [BK06a, BK06b]. All tests were performed on a dual 3-GHz Pentium Xeon desktop computer with 2-GB of RAM.

8.6.1 PARAFAC

We computed a 25-component ($r = 25$) decomposition of the term-author-month array $\mathcal{X}$ using PARAFAC. One ALS iteration of PARAFAC took about 22.5 seconds, requiring an average of 27 iterations to satisfy a tolerance of 10^{-4} in the change of fit. We chose the smallest minimizer from among ten runs starting from random initializations. The relative norm of the difference was 0.8904.

We also computed a 25-component ($r = 25$) decomposition of the term-author-day array $\mathcal{Y}$ using PARAFAC. For this larger dataset, one ALS iteration of PARAFAC took 26 seconds, requiring an average of 13 iterations to satisfy a tolerance of 10^{-4} in the change of fit. We chose the smallest minimizer from among ten runs starting from random initializations. The relative norm of the difference was 0.9562.

8.6.2 Nonnegative Tensor Decomposition

We computed a 25-component ($r = 25$) nonnegative decomposition of the term-author-month array $\mathcal{X}$. One iteration took about 22 seconds, and most runs required less than 50 iterations to satisfy a tolerance of 10^{-4} in the relative change of fit. We chose the smallest minimizer from among ten runs from random starting points, and the relative norm of the difference was 0.8931.

We also computed a 25-component ($r = 25$) nonnegative decomposition of the larger term-author-day array $\mathcal{Y}$. One iteration took about 26 seconds, and the average run required about 17 iterations to satisfy a tolerance of 10^{-4} in the relative change of fit. We chose the smallest minimizer from among ten runs from random starting points, and the relative norm of the difference was 0.9561.

8.6.3 Analysis of Results

PARAFAC is able to identify and track discussions over time in each triad $\{A_j, B_j, C_j\}$, for $j = 1, \ldots, r$. A discussion is associated with the topic and primary participants identified in the columns of A and B, respectively, and the corresponding column of C provides a profile over time, showing the relative activity of that discussion over 12 months or over 357 days. Figures 8.2 and 8.3 present a histogram (or Gantt chart) of the monthly activity for each discussion identified by the classical and nonnegative PARAFAC models, respectively. Similarly, Figures 8.4 and 8.5 present line charts of the daily activity for each discussion identified by the classical and nonnegative PARAFAC models, respectively. We first discuss the monthly results, and then describe the finer grained details uncovered in the daily results.

Fig. 8.2. Six distinguishable discussions among the 25 extracted by classic PARAFAC. Diagonal shading of cells is used to indicate negative components in the tensor groups.

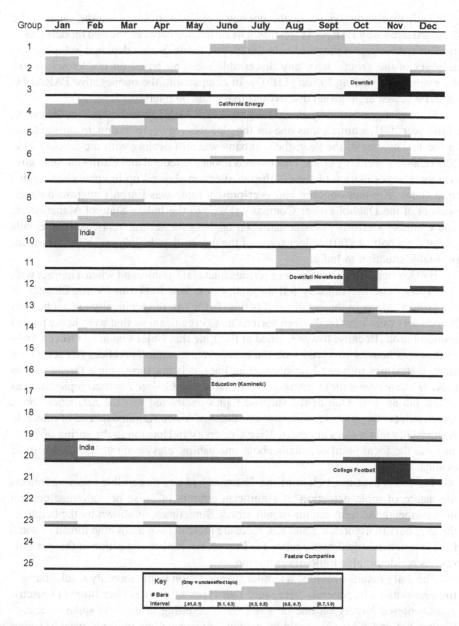

Fig. 8.3. Eight distinguishable discussions among the 25 extracted by nonnegative PARAFAC.

Qualitatively, the results of the nonnegative decomposition are the same as the standard three-way PARAFAC results. The difference between the two models was in the ability to interpret the results. In the 25 discussion groups depicted in Figure 8.2, only six of the groups have any discernible meaning based on our knowledge of the events surrounding Enron [ME03]. In comparison, the nonnegative PARAFAC model revealed eight group discussions that could be interpreted.

The topics generated by the nonnegative PARAFAC model do reflect the events of the year 2001, a tumultuous one for the global energy corporation, to say the least. In the first quarter of the year, the company was still dealing with the fallout of the 2000 California energy crisis. Discussions about the federal and California state governments' investigation of the California situation showed up in emails during this time frame. Another ongoing and ever-present topic was Enron's attempted development of the Dabhol Power Company (DPC) in the Indian State of Maharashtra. The company's efforts in India had been ongoing for several years, and the emails of the early half of 2001 reflect some of the day-to-day dealings with the less-than-profitable situation in India.

By October 2001, Enron was in serious financial trouble, and when a merger with the Dynegy energy company fell through, Enron was forced to file for Chapter 11 bankruptcy. Many of the emails of this time frame (more specifically in October and November) were newsfeeds from various news organizations that were being passed around Enron. Because it was learned at this time that Chief Financial Officer Andy Fastow was heavily involved with the deceptive accounting practices (by setting up sham companies to boost Enron's bottom line), it is not surprising a thread on this topic (*Fastow companies*) emerged. Predictably, the *College Football* topic emerges in late fall as well. One of the surprise topics uncovered was the *Education* topic, which reflects the interests and responsibilities of Vince Kaminski, head of research. Kaminski taught a class at nearby Rice University in Houston in the spring of 2001, and was the focal point of emails about internships, class assignments, and résumé evaluation.

The fact that at most eight of the 25 topics had any discernible meaning reflects the nature of topic detection. A significant amount of *noise* or undefined content may permeate the term-author-month arrays. Sometimes, as shown by the height of the gray bars in Figures 8.2 and 8.3, there are indicators of a possible thread of some kind (not necessarily directly related to Enron), but further inspection of those emails reveals no identifiable topic of discussion.

The daily results provided a similar interpretation as the monthly results but at a finer resolution. In general, there were four types of profiles over time: (1) discussions centered largely on one or a few days, resulting in a single spike (see, e.g., Figures 8.4 and 8.5); (2) continual activity, represented as multiple weekly spikes throughout the year; (3) continual activity with lulls, where a period of calm separates bursts of discussion; and (4) a series of weekly spikes usually spanning three or more months.

Of the 25 groups found with the PARAFAC model, roughly half were single spikes. We have identified three of these groups of particular significance in Figure 8.4. The first identifies a flood of emails about the possible Dynegy/Enron merger (November 11 and 12), which was something new. This merger topic was found in [BB05a], but this topic did not show up previously in our monthly results. The second is a topic on January 7 that the Enron employees (Kean, Hughes, and Ambler) were discussing India based on an article published by Reuters and another media report that required some attention by the Enron executives. The third group also matches what happened on August 27, when discussion was centered on the U.S. Court of Appeals ruling on section 126 of an Environmental Protection Agency code. This specific topic concerning compliance of emission limitations had not yet surfaced anywhere before in our analysis.

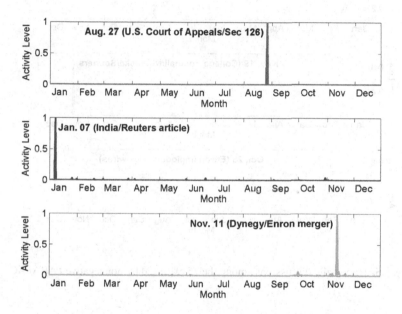

Fig. 8.4. Three daily discussions among the 25 extracted by PARAFAC.

The nonnegative PARAFAC model identified the same temporal patterns as PARAFAC, with a majority being a series of weekly spikes spanning three or more months. Roughly one third are single-spike patterns, and just two discussions are of the bimodal type with a lull. We mention three of the more interesting groups from single-spike discussions, shown in Figure 8.5. The first group of the nonnegative model is interesting because it either reveals a potentially new topic or a specific occurrence of a general topic. On August 22, there was a flurry of emails and responses to an Email called *California Campaign Closeout*. Richard Shapiro praised the employees working in California and many responded to his acknowledgment. A second group is a very strong cluster pointing to college football with terms like Nebraska, Sooners, bowl, Cougars, and Tennessee. These were sent by Matthew Motley on November 20. Finally, a third group was influenced heavily by Enron's final implosion in and around October 25 and 26. There were many news wire stories about the plight of Enron during this time. PARAFAC had found this topic but had identified the time period 2 days earlier. We believe this difference is a result of the random initialization that both were provided.

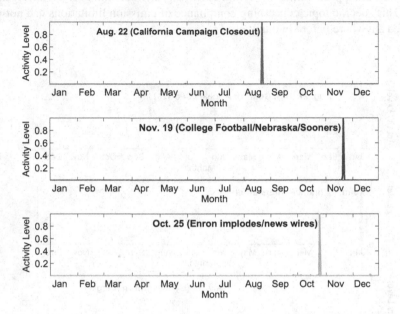

Fig. 8.5. Three daily discussions among the 25 extracted by nonnegative PARAFAC.

8.7 Future Work

As demonstrated by this study, nonnegative tensor factorization (implemented by PARAFAC) can be used to extract meaningful discussions from email communications. The ability to assess term-to-author (or term-to-email) associations both semantically and temporally via three-way decompositions is an important advancement in email surveillance research. Previously reported clusters of Enron emails using nonnegative matrix factorization (i.e., two-way decompositions) [BB05b] [SBPP06, BBL+07] were unable to extract discussions such as the *Education* thread mentioned in Section 8.6.2 or sequence the discussion of the company's downfall by source (news feeds versus employee-generated). The optimal segmentation of *time* as the third dimension for email clustering may be problematic. Grouping emails by month may not be sufficient for some applications, and so more research in the cost-benefit trade-offs of finer time segmentation (e.g., grouping by weeks, days, or even minutes) is needed. Determining the optimal tensor rank r for models such as PARAFAC is certainly another important research topic. Term weighting in three-way arrays is also an area that greatly influences the quality of results but is not yet well understood.

Acknowledgments

This research was sponsored by the United States Department of Energy and by Sandia National Laboratory, a multiprogram laboratory operated by Sandia Corporation, a Lockheed Martin Company, for the United States Department of Energy under contract DE–AC04–94AL85000.

References

[AÇKY05] E. Acar, S.A. Çamtepe, M.S. Krishnamoorthy, and B. Yener. Modeling and multiway analysis of chatroom tensors. In *ISI 2005: IEEE International Conference on Intelligence and Security Informatics*, volume 3495 of *Lecture Notes in Computer Science*, pages 256–268. Springer, New York, 2005.

[BB05a] M.W. Berry and M. Browne. Email surveillance using non-negative matrix factorization. In *Workshop on Link Analysis, Counterterrorism and Security, SIAM Conf. on Data Mining*, Newport Beach, CA, 2005.

[BB05b] M.W. Berry and M. Browne. Email surveillance using nonnegative matrix factorization. *Computational & Mathematical Organization Theory*, 11:249–264, 2005.

[BB05c] M.W. Berry and M. Browne. *Understanding Search Engines: Mathematical Modeling and Text Retrieval*. SIAM, Philadelphia, second edition, 2005.

[BBL+07] M.W. Berry, M. Browne, A.N. Langville, V.P. Pauca, and R.J. Plemmons. Algorithms and applications for approximate nonnegative matrix factorization. *Computational Statistics & Data Analysis*, 52(1):155–173, 2007.

[BK06a] B.W. Bader and T.G. Kolda. Efficient MATLAB computations with sparse and factored tensors. Technical Report SAND2006-7592, Sandia National Laboratories, Albuquerque, New Mexico and Livermore, California, December 2006. Available from World Wide Web: http://csmr.ca.sandia.gov/~tgkolda/pubs.html#SAND2006-7592.

[BK06b] B.W. Bader and T.G. Kolda. MATLAB Tensor Toolbox, version 2.1. http://csmr.ca.sandia.gov/~tgkolda/TensorToolbox/, December 2006.

[CC70] J.D. Carroll and J.J. Chang. Analysis of individual differences in multidimensional scaling via an N-way generalization of 'Eckart-Young' decomposition. *Psychometrika*, 35:283–319, 1970.

[Coh] W.W. Cohen. Enron email dataset. Web page. http://www.cs.cmu.edu/~enron/.

[FBH03] N.M. Faber, R. Bro, and P.K. Hopke. Recent developments in CANDECOMP/PARAFAC algorithms: a critical review. *Chemometr. Intell. Lab. Syst.*, 65(1):119–137, January 2003.

[Fed]   Federal Energy Regulatory Commission. FERC: Information released in En-
             ron investigation. http://www.ferc.gov/industries/electric/
             indus-act/wec/enron/info-release.asp.
[Gri03]   T. Grieve. The decline and fall of the Enron empire. *Slate*, October 14 2003. Avail-
             able from World Wide Web: http://www.salon.com/news/feature/
             2003/10/14/enron/index_np.html.
[GWB03]   J.T. Giles, L. Wo, and M.W. Berry. GTP (General Text Parser) Software for Text
             Mining. In H. Bozdogan, editor, *Software for Text Mining, in Statistical Data
             Mining and Knowledge Discovery*, pages 455–471. CRC Press, Boca Raton, FL,
             2003.
[Har70]   R.A. Harshman. Foundations of the PARAFAC procedure: models and con-
             ditions for an "explanatory" multi-modal factor analysis. *UCLA Working Pa-
             pers in Phonetics*, 16:1–84, 1970. Available at http://publish.uwo.ca/
             ~harshman/wpppfac0.pdf.
[KB06]   T.G. Kolda and B.W. Bader. The TOPHITS model for higher-order web
             link analysis. In *Workshop on Link Analysis, Counterterrorism and Security*,
             2006. Available from World Wide Web: http://www.cs.rit.edu/~amt/
             linkanalysis06/accepted/21.pdf.
[KBK05]   T.G. Kolda, B.W. Bader, and J.P. Kenny. Higher-order web link analysis using
             multilinear algebra. In *ICDM 2005: Proceedings of the 5th IEEE International
             Conference on Data Mining*, pages 242–249. IEEE Computer Society, Los Alami-
             tos, CA, 2005.
[LS99]   D. Lee and H. Seung. Learning the parts of objects by non-negative matrix factor-
             ization. *Nature*, 401:788–791, 1999.
[ME03]   B. Mclean and P. Elkind. *The Smartest Guys in the Room: The Amazing Rise and
             Scandalous Fall of Enron*. Portfolio, New York, 2003.
[MHA06]   M. Mørup, L. K. Hansen, and S. M. Arnfred. Sparse higher order non-negative
             matrix factorization. *Neural Computation*, 2006. Submitted.
[Mør05]   M. Mørup. Decomposing event related eeg using parallel factor (parafac). Presen-
             tation, August 29 2005. Workshop on Tensor Decompositions and Applications,
             CIRM, Luminy, Marseille, France.
[PCMP06]   C.E. Priebe, J.M. Conroy, D.J. Marchette, and Y. Park. Enron dataset. Web page,
             February 2006. http://cis.jhu.edu/~parky/Enron/enron.html.
[SA05]   J. Shetty and J. Adibi. Ex employee status report. Online, 2005. http:www.
             isi.edu/~adibi/Enron/Enron_Employee_Status.xls.
[SBG04]   A. Smilde, R. Bro, and P. Geladi. *Multi-Way Analysis: Applications in
             the Chemical Sciences*. Wiley, West Sussex, England, 2004. Available from
             World Wide Web: http://www.wiley.com/WileyCDA/WileyTitle/
             productCd-0471986917.html.
[SBPP06]   F. Shahnaz, M.W. Berry, V.P. Pauca, and R.J. Plemmons. Document clustering
             using non-negative matrix factorization. *Information Processing & Management*,
             42(2):373–386, 2006.
[SGB00]   N.D. Sidiropoulos, G.B. Giannakis, and R. Bro. Blind PARAFAC receivers for
             DS-CDMA systems. *IEEE Transactions on Signal Processing*, 48(3):810–823,
             2000.
[SZL+05]   J.-T. Sun, H.-J. Zeng, H. Liu, Y. Lu, and Z. Chen. CubeSVD: a novel approach to
             personalized Web search. In *WWW 2005: Proceedings of the 14th International
             Conference on World Wide Web*, pages 382–390. ACM Press, New York, 2005.
[TB05]   G. Tomasi and R. Bro. PARAFAC and missing values. *Chemometr. Intell. Lab.
             Syst.*, 75(2):163–180, February 2005.

[Tuc66]   L.R. Tucker. Some mathematical notes on three-mode factor analysis. *Psychometrika*, 31:279–311, 1966.

9

Spam Filtering Based on Latent Semantic Indexing

Wilfried N. Gansterer, Andreas G.K. Janecek, and Robert Neumayer

Overview

In this chapter, the classification performance of latent semantic indexing (LSI) applied to the task of detecting and filtering unsolicited bulk or commercial email (UBE, UCE, commonly called "spam") is studied. Comparisons to the simple vector space model (VSM) and to the extremely widespread, de-facto standard for spam filtering, the SpamAssassin system, are summarized. It is shown that VSM and LSI achieve significantly better classification results than SpamAssassin.

Obviously, the classification performance achieved in this special application context strongly depends on the feature sets used. Consequently, the various classification methods are also compared using two different feature sets: (1) a set of purely textual features of email messages that are based on standard word- and token-extraction techniques, and (2) a set of application-specific "meta features" of email messages as extracted by the SpamAssassin system. It is illustrated that the latter tends to achieve consistently better classification results.

A third central aspect discussed in this chapter is the issue of problem reduction in order to reduce the computational effort for classification, which is of particular importance in the context of time-critical on-line spam filtering. In particular, the effects of truncation of the SVD in LSI and of a reduction of the underlying feature set are investigated and compared. It is shown that a surprisingly large amount of problem reduction is often possible in the context of spam filtering without heavy loss in classification performance.

9.1 Introduction

Unsolicited bulk or commercial email (UBE, UCE, "spam") has been a severe problem on the Internet in the last several years. Although many strategies for addressing this problem have been proposed, we are still far away from a satisfactory and fundamental solution. In part, this is due to the fact that many of the methods proposed and developed into concepts usable in practice have an ad-hoc nature and lack a lasting effect. Those individuals who make a profit based on the spam phenomenon (the "spammers") can easily find ways to circumvent the proposed methods.

The problem of filtering spam is a binary classification problem in the sense that every incoming email has to be classified as "spam" or "not spam." The investigations summarized here were motivated by the general assumption that spam email messages tend to have several semantic elements in common, which are usually not present in regular email (commonly called "ham"). This assumption is plausible due to the economic aspects underlying the spam phenomenon (see, for example, [ISGP06] for a discussion). However, the common semantic elements are hard to pinpoint comprehensively, because they are not fully known, not always explicit, or may change dynamically. In other words, they are in some sense implicit or *latent*. Some approaches that were developed have tried to concentrate on some specific properties of (current) spam and to design anti-spam methods based thereon. However, due to the difficulties in identifying a comprehensive set of properties that unambiguously characterize spam, these approaches have not led to fundamental and persistent anti-spam strategies.

9.1.1 Approaches Investigated

In this chapter, the classification performance of a *vector space model* (VSM) and *latent semantic indexing* (LSI) applied to the task of spam filtering are studied. Based on a feature set used in the extremely widespread, de-facto standard spam filtering system known as SpamAssassin and also on subsets selected from this feature set, a vector space model and latent semantic indexing are applied for classifying email messages as spam or not spam. The test datasets used are partly from the official TREC 2005 dataset and partly self-collected.

The investigation of LSI for spam filtering summarized here evaluates the relationship between two central aspects: (1) the truncation of the SVD in LSI, and (2) the resulting classification performance in this specific application context. It is shown that a surprisingly large amount of truncation is often possible without heavy loss in classification performance. This forms the basis for good and extremely fast approximate (pre-) classification strategies, which are very useful in practice.

The approaches investigated in this chapter are shown to compare favorably to two important alternatives, in that they achieve better classification results than SpamAssassin, and they are better and more robust than a related LSI-based approach using textual features, which has been proposed earlier.

The low-rank approximation of the feature-document matrix within LSI is one possibility to reduce the potentially huge size of the classification problem. As an alternative, we investigate strategies for reducing the feature set *before* the classification process. Compared to the LSI truncation process on a full feature set, this approach can potentially reduce the computing time as the matrices in the time-consuming SVD computation become smaller. The classification performances of these two approaches to problem size reduction are investigated and compared.

9.1.2 Background and Related Work

Existing anti-spam methods can be categorized according to their point of action in the email transfer process. Consequently, three classes of approaches can be distinguished (for details, see [GIL+05]). *Pre-send* methods act before the email is transported over the network, whereas *post-send* methods act after the email has been transferred to the receiver. A third class of approaches comprises new protocols, which are based on modifying the transfer process itself. Pre-send methods (see, for example, [Bac02] or [GHI+05]) are very important because of their potential to avoid the wasting of resources (network traffic, etc.) caused by spam. However, the efficiency of pre-send methods and of new protocols heavily depends on their widespread deployment. It is unrealistic to expect global acceptance and widespread use of these new methods in the near future, and thus the third group of methods, post-send spam filtering methods, will continue to be the "workhorse" in spam defense.

Many of the spam filtering concepts currently used in practice are mostly static in nature, such as black- and whitelisting or rule-based filters. A de-facto standard of a rule-based spam filtering system is SpamAssassin [Apa06]. SpamAssassin extracts a multitude of features from incoming email messages, comprising pieces of information from header, body, and the full text of the message. One of the tests integrated into SpamAssassin is a Bayesian classifier, a statistical approach exclusively based on textual features. An example of its application to the scenario of spam filtering is given in [AKCS00]. The underlying idea is to compute a conditional probability for an email being spam based on the words (*tokens*) it contains. From the point of view of the individual user, these and similar filtering methods may achieve reasonably satisfactory results provided they are trained, tuned, and maintained permanently. The user effort required for sustaining satisfactory performance is one of the disadvantages of almost all existing filtering methods.

The application of state-of-the-art information retrieval and text-mining techniques to the tasks of spam filtering and email classification is a topic of current research. An approach closely related to the methods investigated in this chapter has been proposed by Gee [Gee03], where an LSI-based classifier is used for spam filtering. However, in contrast to the investigations summarized here, Gee's approach is exclusively based on *textual* features of the email messages. One of the main motivations for our work was to investigate the influence of the choice of different feature sets on the performance of VSM- and LSI-based spam filtering. In particular, our objective was to extend Gee's approach (LSI on text-based features), to evaluate the classification performance achieved with LSI on a set of "meta features," such as those used in SpamAssassin, and to quantify the influence of feature set reductions on the classification results.

9.1.3 Synopsis

In Section 9.2, the basic principle of LSI is reviewed briefly, and in Section 9.3 its adaptation to the context of spam filtering is discussed. Section 9.4 describes our implementation of this concept and summarizes a set of experiments performed with two email test sets. In Section 9.5, conclusions are drawn and future work in this context is outlined.

9.2 Classification Methodology

In this section, the standard VSM and LSI methods—as, for example, used in Web search engines [LM04]—are reviewed briefly. Their adaptation and application to the task of spam filtering is discussed in Section 9.3.

9.2.1 The Basic Vector Space Model

In information retrieval, a vector space model (VSM) is widely used for representing information. Documents and queries are represented as vectors in a potentially very high dimensional metric vector space. The distance between a query vector and the document vectors is the basis for the information retrieval process, which we summarize here very briefly in its standard form (for more details see, for example, [Lan05]).

Generally speaking, a vector space model for n documents is based on a certain set F of f features, $F = \{F_1, F_2, \ldots, F_f\}$. A feature extraction component has to be available to extract the values v_{ij} for each of these features F_i from each document j. The f-dimensional (column) vector V_j with the components $v_{ij}, i = 1, 2, \ldots, f$, then represents document j in the VSM. The $f \times n$ *feature-document matrix* M (in other contexts called the *term-document matrix*) is composed using all the vectors V_j as columns.

Given a query vector q of length f, the distances of q to all documents represented in M can then be measured (for example) in terms of the cosines of the angles between q and the columns of M. The column with the smallest angle (largest cosine value) represents the closest match between the document collection and the query vector.

9.2.2 LSI

Latent semantic indexing (LSI) or latent semantic analysis (LSA) is a variant of the basic vector space model. Instead of the original feature-document matrix M it operates on a low-rank approximation to M. More specifically, based on the singular value decomposition of M, a low-rank approximation M_k is constructed:

$$M = U \Sigma V^\top \approx U_k \Sigma_k V_k^\top =: M_k, \tag{9.1}$$

where Σ_k contains the k largest singular values of M and $k < f$. Technically, this approximation replaces the original matrix M by another matrix whose column space is a subspace of the column space of M.

LSI Truncation

We call the approximation process (9.1) *LSI truncation* and denote the amount of truncation as a percentage value: If Σ_k contains only those k singular values of M that are greater than p percent of the maximum singular value of M, then the approximation is called "LSI $p\%$" truncation. If, for example, the maximum singular value of M is 100, then for "LSI 2.5%" all k singular values that are greater than 2.5 are used in the approximation of M. Note that as the truncation parameter p increases, the rank of the approximation matrix decreases.

One of the main drawbacks of LSI is the rather large (sometimes even prohibitive) computational cost required for computing the singular value decomposition. On the other hand, it has advantages in terms of storage requirements and computational cost for determining the distances between the query vector q and the documents. These and other considerations are discussed in greater detail in [BDJ99].

9.3 VSM and LSI for Spam Filtering

The main focus of this section is the application of VSM and LSI to the task of spam filtering. In this context, each document is an email message that is represented by a vector in the vector space model.

To construct this vector space model, each email message has to be represented by a set of feature values determined by a feature extraction component. This leads to two central questions arising in the context of spam filtering methodology: (1) Which features of email data are (most) relevant for the classification into spam and ham (textual features, header information, etc.)? (2) Based on a certain set of features, what is the *best* method for categorizing email messages into spam and ham? In the remainder of this section, we will mainly focus on the former question. The evaluation of VSM and LSI summarized in Section 9.4 is an attempt to contribute to answering the latter question.

9.3.1 Feature Sets

In this study, we consider and compare two types of features that are very important and widely used in the context of spam filtering. On the one hand, we use the features extracted by the state-of-the-art spam filtering system SpamAssassin. This feature set is denoted by "F_SA" in the following. On the other hand, we use a comparable number of purely text-based features. This feature set is denoted by "F_TB" in the following. First, these two types of feature sets and their extraction are discussed. Then, we discuss feature/attribute selection methods for investigating a controlled *reduction* of the number of features in each set.

SpamAssassin Features

For our first feature set we use the SpamAssassin system [Apa06] to extract their values from each message. SpamAssassin applies many tests and rules (795 in the version used) to extract these features from each email message. Different parts of each email message are tested, comprising the message body (56% of all tests) and the message header (36% of the tests). Moreover, there are URI checks (5% of the tests) as well as rawbody tests (multimedia email (MIME) checks, 2% of the tests) and blacklist tests (1% of the tests). Tests acting on the message body comprise Bayesian classifiers, language-specific tests, tests searching for obfuscated words, html tests, and many others (for a complete list, see `http://spamassassin.apache.org/tests\_3\_1\_x.html`).

Each test is assigned a certain value. This value is a positive real number if the test points to a spam message, and a negative real number if the test points to a ham message. The specific values have been derived by the SpamAssassin team using a neural network trained with error back propagation. In the SpamAssassin system, the overall rating of a message is computed by summing up all values assigned to the tests. If this sum exceeds a user-defined threshold (the default threshold is set to 5), a message is classified as spam. Increasing the threshold will decrease the spam detection rate but will also reduce the false-positive rate, whereas decreasing it will increase the spam detection rate but also the false-positive rate.

Although the number of features extracted from each email message by the SpamAssassin system is very large, experimental analysis shows that only a relatively small subset of these features provides widely applicable useful information. More specifically, for our datasets (see Section 9.4.1) only about *half* of all Spam-Assassin tests triggered at least once, and only about 4% of the tests triggered for more than 10% of the messages. Besides the Bayesian classifier (which returns a value for every message) the SpamAssassin tests triggering most often for our datasets are the *"all_trusted"* test, which checks whether a message has only passed through trusted SMTP relays (this test points to ham), as well as the *"razor2"* real-time blacklist test (this test points to spam). A more detailed list of the tests triggering most often for the datasets used in this work can be found in [GIJ+06].

It should be mentioned that the SpamAssassin feature set is *"unbalanced"* in the sense that about 96% of all tests check for a feature that points to spam and only about 4% of the tests point to ham. Another potential problem is that some of the SpamAssassin tests tend to trigger incorrectly. For our test data, 11 tests triggered wrongly for more than 2% of the messages, and seven of them triggered wrongly more often than correctly (see [GIJ+06]). Obviously, this tends to have a negative impact on the classification performance achieved. However, most of these problematic tests are assigned only a low default score, and therefore their impact on the overall classification result is not too strong in practice.

In the work summarized here, we selected those 377 and 299 SpamAssassin features, respectively, which triggered at least once for both of the datasets used (see Section 9.4.1) as a first starting point. We used the features in binary form and also the original values assigned by SpamAssassin. In some cases, using binary features, turned out to yield only slightly better classification results than using the original, weighted values. The experimental results shown in Section 9.4 are based on binary feature values. The shortcomings of the SpamAssassin feature set mentioned above motivated our investigation of improved feature selection and extraction strategies summarized later in this section.

Text-Based Features

The classic alternative, which is used widely, not only in text mining but also in the area of spam filtering, is a purely text-based feature set. Consequently, we consider a second feature set, which consists of words extracted from the email messages. *Document frequency thresholding* is used for reducing the potentially prohibitive dimensionality.

Dimensionality Reduction

When tokenizing text documents, one often faces very high dimensional data. Tens of thousands of features are not easy to handle; therefore, feature selection plays a significant role. Document frequency thresholding achieves reductions in dimensionality by excluding terms having very high or very low document frequencies. Terms (words, tokens) that occur in almost all documents in a collection do not provide any discriminating information. A good overview of various term selection methods explaining their advantages and disadvantages is given in [YP97].

It is usually possible to scale down the dimensionality to about 300 features. Document frequency thresholding is often used as a first step to reduce the dimensionality. Furthermore, techniques such as *information gain* can be used to select the most important features from that "pre-selection" in order to further reduce computational cost.

Document Frequency Thresholding

Document frequency thresholding is even feasible for unsupervised data. The basic assumption is that very frequent terms are less discriminative when trying to distinguish between classes (a term occurring in every single spam and ham message would not contribute to differentiate between them). The majority of tokens (words), however, tends to occur only in a very small number of documents. The biggest advantage of document frequency thresholding is that class information is not needed. It is therefore mainly used for clustering applications where the data consists only of one class or where no class information is available at all. Besides, document frequency thresholding is relatively inexpensive in terms of computational cost. In the context considered here it can be used for dimensionality reduction, for clustering, and to compare the classification results obtained by more sophisticated approaches. Document frequency thresholding proceeds as follows:

- First, the upper threshold is fixed to 0.5, hence all terms that occur in more than half of the documents are omitted.
- Then, the lower threshold is dynamically adapted in order to achieve a predefined desired number of features.

Information Gain

Information gain (IG) is a technique originally used to compute splitting criteria for decision trees. The basic idea behind IG is to find out how well each single feature separates the given dataset.

The overall entropy I for a given dataset S is computed according to the following equation:

$$I = - \sum_{i=1}^{C} p_i \log_2 p_i, \qquad (9.2)$$

where C denotes the total number of classes and p_i the portion of instances that belong to class i. Now, the reduction in entropy or gain in information is computed for each attribute A or token according to

$$IG(S, A) = I(S) - \sum_{v \in A} \frac{|S_v|}{|S|} I(S_v), \qquad (9.3)$$

where v is a value of A and S_v is the number of instances where A has that value v (i.e., the number of instances A occurs in).

According to Eq. (9.3), an information gain value can be computed for each token extracted from a given document collection. Documents are then represented by a given number of tokens having the highest information gain values.

Feature/Attribute Selection Methods Used

We used various feature/attribute selection methods to reduce the number of features within both feature sets introduced before. On that basis, we create different feature subsets comprising only the top discriminating features for both feature sets. This feature reduction reduces the computational effort in the classification process for each message and therefore saves computing time.

The features within the text-based feature set are already ranked based on information gain. We applied information gain attribute selection to the SpamAssassin feature set as well. The computation of the information gain for each SpamAssassin feature can be performed similarly to the computation of the information gain for each token for the text-based features (cf. Eq. (9.2) and (9.3)). Based on the resulting ranking of features, two subsets containing the top 50 and the top 10 features, respectively, were selected for both feature sets, text-based and SpamAssassin features.

For the SpamAssassin feature set, we used two more feature selection strategies for comparison purposes—χ^2 attribute selection and an intuitive strategy: (1) The χ^2 attribute selection evaluates the significance of an attribute by computing the value of the χ^2 statistic with respect to the class. (2) As an intuitive strategy, we extract two feature subsets containing only the top 50 and the top 10 triggering features, respectively (i.e., those SpamAssassin tests that have triggered most often for all email messages within our training data).

For the computation of the information gain and χ^2 attribute selection we used the publicly available machine learning software WEKA [WEK06]. Applying the

χ^2 attribute selection to the text-based features (which have already been ranked by information gain did) not change the order of the features. When comparing the three feature selection methods for the SpamAssassin feature set, eight SpamAssassin features occur in all three top 10 subsets extracted. These comprise Bayesian classifiers (*"bayes_00"* and *"bayes_99"*), special black- and whitelist tests (*"razor2_cf_range_51_100"* and *"razor2_check"*), tests that check whether the message body contains an URL listed in a special URL blacklist (*"uribl_ws_surbl"* and *"uribl_ob_surbl"*), a test that checks whether all headers in the message were inserted by trusted SMTP relays (*"all_trusted"*), and a test that checks whether the message contains html tags (*"html_message"*).

9.3.2 Training and Classification

The application of VSM and LSI to spam filtering investigated here involves two phases: a training phase and the actual classification phase. The training phase comprises the indexing of two known datasets (one consisting of spams and one consisting of hams) and, in the case of LSI, the computation of the singular value decomposition of the feature- or term-document matrix. The classification phase comprises the query indexing and the retrieval of the closest message from the training sets. A newly arriving message can be classified by indexing it based on the feature set used and comparing the resulting query vector to the vectors in the training matrices. If the closest message is contained in the spam training set, then the query message is classified as spam, otherwise it is classified as ham. The distance measurement is based on the angles between the query vector and the training vectors and is computed analogously to standard VSM (see [Lan05]).

9.3.3 Performance Metrics

In the context of spam filtering, a "positive" denotes an email message that is classified as spam and a "negative" denotes an email message that is classified as ham. Consequently, a "true positive" is a spam message that was (correctly) classified as spam, and a "false positive" is a ham message that was (wrongly) classified as spam.

The main objective is to maximize the rate of "true positives" and to minimize rate of "false positives" simultaneously (in order not to lose ham messages). Sometimes, these two metrics are aggregated into a single one, for example, by computing the overall percentage of correctly classified messages (correctly classified spams plus correctly classified hams) in all messages classified (see Figures 9.3 and 9.6).

9.4 Experimental Evaluation

The concepts and methods discussed before have been implemented and evaluated experimentally in the context of spam filtering. The results achieved are summarized in this section.

9.4.1 Data Sets

For the experiments we used two different datasets, one consisting of a part (25%) of the TREC 2005 spam and ham corpus. This dataset is denoted as "S1" in the following and is publicly available [CL05]. The other dataset consists of self-collected email messages and is denoted as "S2" in the following. The messages in S2 have been collected recently over a period of several months from several sources in order to achieve as much diversity as possible. Spam was mostly collected from various spam traps, and ham came from volunteers (mainly private email). S1 contains 9,751 ham messages and 13,179 spam messages. S2 contains 5,502 ham messages and 5,502 spam messages.

9.4.2 Experimental Setup

We performed two different cross-validations for our datasets, a threefold cross-validation for the larger sample S1 and a tenfold cross-validation for the smaller sample S2. For an n-fold cross-validation, we split each of our samples randomly into n parts of roughly equal size and used alternately $n-1$ of these parts for training and one of them for testing. The true/false-positive rates and the true/false-negative rates were measured and the aggregated classification results were computed from these results. The cross-validations were performed for six different LSI truncations using the feature sets F_SA and F_TB as defined in Section 9.3.

The experiments involved processing a large amount of data and a variety of different parameter settings. We used six different machines for the test runs, ranging from a Sun Sparc 64-bit multiprocessor running SunOS v5.10 to standard desktop PCs running Windows XP or Linux Ubuntu 4.0.3.

9.4.3 Analysis of Data Matrices

The average ranks k of the truncated SVD matrices for both datasets and both feature sets are listed in Tables 9.1 and 9.2.

Table 9.1. Rank k of the truncated SVD matrices for different cut-off values in the singular values for S1

Truncation:	0.0%	2.5%	5.0%	10.0%	25.0%	50.0%
Features F_SA						
kHam:	377	24	12	5	1	1
kSpam:	377	94	47	23	6	1
Features F_TB						
kHam:	377	83	37	14	4	3
kSpam:	377	86	50	19	7	2

Table 9.2. Rank k of the truncated SVD matrices for different cut-off values in the singular values for S2

Truncation:	0.0%	2.5%	5.0%	10.0%	25.0%	50.0%
Features F_SA						
kHam:	299	31	15	7	3	1
kSpam:	299	79	42	17	5	2
Features F_TB						
kHam:	299	53	26	14	4	2
kSpam:	299	55	27	11	4	2

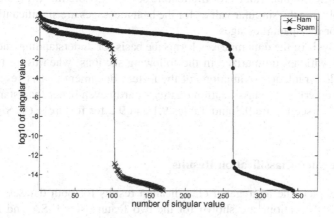

Fig. 9.1. Singular values in vector space models for sample S1 using feature set F_SA.

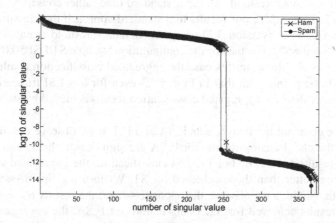

Fig. 9.2. Singular values in vector space models for sample S1 using feature set F_TB.

The distribution of the singular values in the vector space models for sample S1 is illustrated in Figures 9.1 and 9.2. The sharp decline in the magnitude of the singular values for both classes (ham and spam) is very interesting.

Comparing Figures 9.1 and 9.2, it can be observed that the singular values for feature set F_TB tend to be significantly larger than the singular values for feature set F_SA. Until the sharp decline at about singular value number 250 (for both ham and spam features), all singular values for the feature set F_TB are larger than 5, whereas the majority of the singular values for feature set F_SA is much smaller than 5 (93% for the ham sample and 66% for the spam sample, respectively). Moreover, the largest singular values for the feature set F_TB are about 8 to 10 times larger than the largest singular values for the feature set F_SA. Looking at Figure 9.1, it is clearly visible that the singular values for the ham messages are significantly smaller than those for the spam messages.

This analysis of the data matrices forms the basis for understanding and explaining the observations summarized in the following sections: when using feature set F_SA, very low-rank approximations of the feature-document matrix (even $k = 1$) still achieve a very good classification quality—partly even better than standard SpamAssassin (see Figure 9.3 and Tables 9.1 and 9.2 for feature set F_SA and LSI 50%).

9.4.4 Aggregated Classification Results

Figure 9.3 depicts the aggregated classification results for both datasets used. Six different LSI truncations are shown for the two feature sets F_SA and F_TB. As mentioned in Section 9.3.1, SpamAssassin assigns positive values to features pointing to spam and negative values to features pointing to ham. Classification is then performed based on a threshold value, the standard (and rather conservative) default value being 5. To compare our results to a standard approach the classification results of SpamAssassin (version 3.1) using this default threshold are also shown in Figure 9.3 ("SA th=5"). The bar for the configuration [sample S1/LSI 50.0%/feature set F_TB] is not visible, as in this case the aggregated classification result is below 80%. It has to be pointed out that in Figure 9.3 even for low LSI truncations most results are better than the aggregated classification results achieved by standard SpamAssassin.

When we compare the feature sets F_SA and F_TB, we observe that for all LSI truncations the classification results for F_SA are significantly better than the corresponding results for F_TB. For the LSI classification, the aggregated results for sample S2 are better than those achieved for S1. When using SpamAssassin with the default threshold 5 we observe the opposite—here, the results achieved for S1 exceed the results achieved for S2. Using feature set F_SA, the aggregated results for LSI 2.5% and LSI 5.0% are even slightly better than the results for LSI 0.0% (standard VSM), where all features are used for the classification. In contrast, for feature set F_TB almost every increase of the LSI truncation (decrease in the rank of the truncated SVD) causes a decrease in the classification result (being the case [sample S2/LSI 50.0%]).

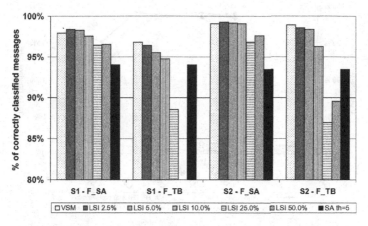

Fig. 9.3. Aggregated classification results for S1 and S2.

9.4.5 True/False-Positive Rates

In contrast to Figure 9.3, Figures 9.4 and 9.5 show true- and false-positive rates separately. The SpamAssassin threshold was again set to 5. Using a more aggressive threshold for SpamAssassin would increase both the false-positive rate *and* the true-positive rate (e. g., for sample S1 a SpamAssassin threshold of 3 achieved a true/false-positive rate of 93.89% and 1.94%, respectively).

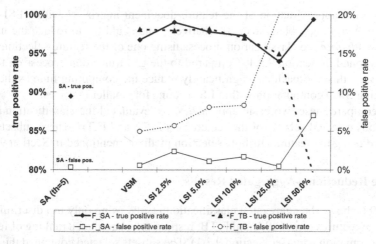

Fig. 9.4. True- and false-positive rates for sample S1.

Both figures show a similar behavior of the true/false-positive rates for F_SA and F_TB and indicate a significantly higher false-positive rate when using feature set F_TB. In particular, for sample S1 the false-positive rate of F_TB is quite

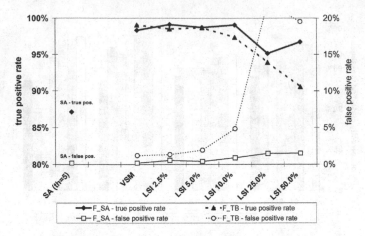

Fig. 9.5. True- and false-positive rates for sample S2.

high (4.91% using VSM), although the aggregated classification result reaches a respectable 96.80% (see Figure 9.3, F_TB using VSM). The false-positive rate for sample S2 using feature set F_TB is slightly lower but still consistently higher than when feature set F_SA is used. The false-positive rates tend to remain almost constant except for high LSI truncations where they increase significantly.

9.4.6 Feature Reduction

The low-rank approximation of the feature-document matrix M within LSI is one way to reduce the problem size. Alternatively, one could try to reduce the number of features *before* the classification process using one of the feature selection methods mentioned in Section 9.3.1. Compared to the LSI truncation process on the full feature sets, this approach can significantly reduce the computing time as the time-consuming SVD computation within LSI is done for smaller matrices.

To compare these two basic approaches, we evaluated the classification results for various selected subsets of the feature sets F_SA and F_TB. These subsets were extracted using the feature/attribute selection methods mentioned in Section 9.3.1.

Feature Reduction—Aggregated Results

Figure 9.6 shows the aggregated classification results using only top discriminating subsets of features from F_SA and F_TB, respectively, based on a ranking of features using information gain (see Section 9.3.1). The subsets selected contained the top 50 and the top 10 discriminating features, respectively. For classification, VSM and LSI 25% were used.

It can be observed that classification results based on a reduced number of features are slightly worse than the results achieved with the original feature sets (except for the top 50 F_TB features for sample S1, where the classification results are

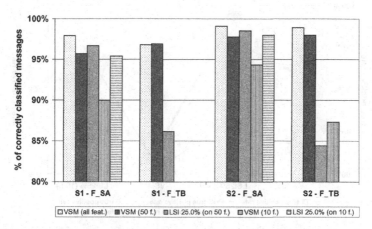

Fig. 9.6. Aggregated results. VSM and LSI classification on reduced feature sets using information gain.

nearly equal). When applying LSI to reduced feature sets, only the results for feature set F_SA remain acceptable. Thus, feature set F_SA turns out to be more robust in terms of feature reduction than feature set F_TB. It is also very interesting to note that the LSI 25% classification based on ten selected features outperforms the VSM classification based on the same 10 features for both samples and both feature sets.

Feature Reduction—True/False-Positive Rates

Figures 9.7 and 9.8 show the true/false-positive rates for samples S1 and S2, respectively, using different subsets of feature set F_SA. Three different attribute selection methods have been applied to extract these subsets (see Section 9.3.1).

While the classification results for the subsets selected by information gain and χ^2 attribute selection are almost equal (except for the true-positive rate for sample S1 in Figure 9.7), the results for the subsets containing the top triggering tests (TT) differ slightly. All curves (except the TT true-positive rate for Figure 9.8) show a similar behavior: When LSI 25% is applied (on the top 50 and top 10 feature subsets, respectively) the true-positive *and* the false-positive rates tend to increase in comparison to the basic VSM. Focusing on the classification results for VSM based on the top 50 and the top 10 features, respectively, it can be observed that for all subsets the false-positive rates are nearly equal, whereas the true-positive rates tend to decrease when the number of features is reduced. Contrary to that, the results for LSI 25% based on 10 features are almost equal to the results of the same LSI truncation based on 50 features. Only for sample S1 the false-positive rate for all three feature selection methods increases slightly when LSI 25% is based on only 10 features.

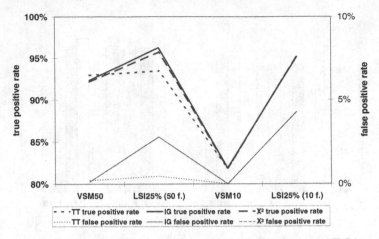

Fig. 9.7. Feature selection results for sample S1 using feature set F_SA.

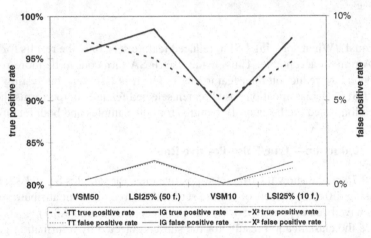

Fig. 9.8. Feature selection results for sample S2 using feature set F_SA.

9.5 Conclusions

In this chapter, the application of latent semantic indexing (LSI) to the task of spam filtering has been investigated. The underlying vector space models are based on two different feature sets: Purely text-based features resulting from word/token extraction (similar to the approach investigated by [Gee03]) were compared with a feature set based on the widespread SpamAssassin system for spam filtering. Moreover, several feature selection methods for extracting subsets of features containing the top discriminating features from each of the feature sets have been investigated as an alternative way of reducing the problem size. Experimental evaluations on two large datasets showed several interesting results:

1. For both feature sets (F_SA and F_TB), VSM and LSI achieve significantly better classification results than the extremely widespread de-facto standard for spam filtering, SpamAssassin.
2. The classification results achieved with both VSM and LSI based on the feature set of SpamAssassin (F_SA) are consistently better than the results achieved with purely textual features (F_FB).
3. For both feature sets, VSM achieves better classification results than LSI at many truncation levels. However, when using LSI based on the SpamAssassin feature set, the classification results are surprisingly robust to LSI truncations with very low rank. This indicates that LSI can provide ways for computing good (approximate) "preclassifications" extremely fast, which is very useful in practice.
4. Distinguishing true- and false-positive rates, we observed that for all truncation levels of LSI the false positive rate, which is a very important performance metric in spam filtering, based on feature set F_SA is much lower than the one based on feature set F_TB. Standard SpamAssassin also achieves a good false-positive rate; however, its true-positive rate is rather poor compared to VSM and LSI.
5. Among the approaches for reducing the problem size and thus the computational effort for the classification problem, LSI truncation based on the original feature sets turned out to be more stable than VSM classification using only subsets of the feature sets. Especially for feature set F_TB, the classification results using only the top 10 discriminating features were poor compared to the LSI 10% classification using low-rank approximations of the original feature-document matrix (between $k = 11$ and $k = 19$; see Tables 9.1 and 9.2).
6. The classification performance of LSI based on SpamAssassin features is also quite robust if the number of features used is reduced significantly and still achieves remarkably good aggregated classification results, which is not the case for LSI based on purely textual features.

Overall, the experiments indicate that VSM and LSI are very well suited for spam filtering if the feature set is properly chosen. In particular, we showed that the Spam-Assassin feature set achieves better results than the commonly used purely textual feature sets based on word- and token-extraction. Both VSM and LSI based on properly chosen small subsets of the SpamAssassin feature set are very well suited as approximate but highly efficient preclassification strategies, which can assign a big part of a live-stream of email messages to a given category very fast (for example, in the context of a recently developed component-based architecture for spam filtering [GJL07]).

Our current work focuses on more analysis of various feature selection and feature extraction strategies in order to further improve upon the currently used Spam-Assassin feature set and on comparisons with other well-established classification methods (see, for example, [CS02]) in the context of spam filtering. Moreover, another important topic in our ongoing work is the utilization of special sparsity structures for classification. Applying various types of sparse matrix techniques will make VSM highly competitive, in particular in terms of computational cost.

Acknowledgments

We would like to express our gratitude to Internet Privatstiftung Austria, mobilkom austria, and UPC Telekabel for supporting this research. We also thank the anonymous referees for their valuable comments, which helped us improve our work.

References

[AKCS00] I. Androutsopoulos, J. Koutsias, K. Chandrinos, and C.D. Spyropoulos. An experimental comparison of naive bayesian and keyword-based anti-spam filtering with personal e-mail messages. In *SIGIR '00: Proceedings of the 23rd Annual International ACM SIGIR Conference on Research and Development in Information Retrieval*, pages 160–167, 2000. Available from World Wide Web: http://doi.acm.org/10.1145/345508.345569.

[Apa06] Apache Software Foundation. SpamAssassin open-source spam filter, 2006. Available from World Wide Web: http://spamassassin.apache.org/.

[Bac02] A. Back. Hashcash—a denial of service counter-measure, 2002. Available from World Wide Web: http://www.hashcash.org/papers/hashcash.pdf.

[BDJ99] M.W. Berry, Z. Drmac, and E.R. Jessup. Matrices, vector spaces, and information retrieval. *SIAM Review*, 41(2):335–362, 1999.

[CL05] G.V. Cormack and T.R. Lynam. TREC 2005 spam public corpora, 2005. Available from World Wide Web: http://plg.uwaterloo.ca/cgi-bin/cgiwrap/gvcormac/foo.

[CS02] N. Cristiani and B. Scholkopf. Support vector machines and kernel methods: the new generation of learning machines. *AI Magazine*, 23:31–41, 2002.

[Gee03] K.R. Gee. Using latent semantic indexing to filter spam. In *ACM Symposium on Applied Computing, Data Mining Track*, pages 460–464, 2003. Available from World Wide Web: http://ranger.uta.edu/~cook/pubs/sac03.ps.

[GHI+05] W.N. Gansterer, H. Hlavacs, M. Ilger, P. Lechner, and J. Strauß. Token buckets for outgoing spam prevention. In M.H. Hamza, editor, *Proceedings of the IASTED International Conference on Communication, Network, and Information Security (CNIS 2005)*. ACTA Press, Anaheim, CA, November 2005.

[GIJ+06] W.N. Gansterer, M. Ilger, A. Janecek, P. Lechner, and J. Strauß. Final report project 'Spamabwehr II'. Technical Report FA384018-5, Institute of Distributed and Multimedia Systems, Faculty of Computer Science, University of Vienna, 05/2006.

[GIL+05] W.N. Gansterer, M. Ilger, P. Lechner, R. Neumayer, and J. Strauß. Anti-spam methods—state of the art. Technical Report FA384018-1, Institute of Distributed and Multimedia Systems, Faculty of Computer Science, University of Vienna, March 2005.

[GJL07] W.N. Gansterer, A.G.K. Janecek, and P. Lechner. A reliable component-based architecture for e-mail filtering. In *Proceedings of the Second International Conference on Availability, Reliability and Security (ARES 2007)*, pages 43–50. IEEE Computer Society Press, Los Alamitos, CA, 2007.

[ISGP06] M. Ilger, J. Strauß, W.N. Gansterer, and C. Proschinger. The economy of spam. Technical Report FA384018-6, Institute of Distributed and Multimedia Systems, Faculty of Computer Science, University of Vienna, September 2006.

[Lan05] A.N. Langville. The linear algebra behind search engines. In *Journal of Online Mathematics and Its Applications (JOMA), 2005, Online Module*, 2005. Available from World Wide Web: http://mathdl.maa.org/mathDL/4/?pa=content&sa=viewDocument&nodeId=636.

[LM04] A.N. Langville and C.D. Meyer. The use of linear algebra by web search engines. *IMAGE Newsletter, 33:2-6*, 2004.

[WEK06] WEKA, 2006. Available from World Wide Web: http://www.cs.waikato.ac.nz/ml/weka/. Data Mining Software in Java.

[YP97] Y. Yang and J.O. Pedersen. A comparative study on feature selection in text categorization. In Douglas H. Fisher, editor, *Proceedings of ICML-97, 14th International Conference on Machine Learning*, pages 412–420, Nashville, TN, 1997. Morgan Kaufmann Publishers, San Francisco. Available from World Wide Web: citeseer.ist.psu.edu/yang97comparative.html.

Part IV

Anomaly Detection

10

A Probabilistic Model for Fast and Confident Categorization of Textual Documents

Cyril Goutte

Overview

We describe the National Research Council's (NRC) entry in the Anomaly Detection/Text Mining competition organized at the Text Mining 2007 Workshop[1] (see Appendix). This entry relies on a straightforward implementation of a probabilistic categorizer described earlier [GGPC02]. This categorizer is adapted to handle multiple labeling and a piecewise-linear confidence estimation layer is added to provide an estimate of the labeling confidence. This technique achieves a score of 1.689 on the test data. This model has potentially useful features and extensions such as the use of a category-specific decision layer or the extraction of descriptive category keywords from the probabilistic profile.

10.1 Introduction

We relied on the implementation of a previously described probabilistic categorizer [GGPC02]. One of its desirable features is that the training phase is extremely fast, requiring only a single pass over the data to compute the summary statistics used to estimate the parameters of the model. Prediction requires the use of an iterative maximum likelihood technique (expectation maximization (EM) [DLR77]) to compute the posterior probability that each document belongs to each category.

Another attractive feature of the model is that the probabilistic profiles associated with each class may be used to describe the content of the categories. Indeed, even though, by definition, class labels are known in a categorization task, those labels may not be descriptive. This was the case for the contest data, for which only the category number was given. It is also the case, for example, with some patent classification systems (e.g., a patent on text categorization may end up classified as "G06F 15/30" in the international patent classification, or "707/4" in the U.S.).

---

[1] http://www.cs.utk.edu/tmw07

In the following section, we describe the probabilistic model, the training phase, and the prediction phase. We also address the problem of providing multiple labels per documents, as opposed to assigning each document to a single category. We also discuss the issue of providing a confidence measure for the predictions and describe the additional layer we used to do that. Section 10.3 describes the experimental results obtained on the competition data. We provide a brief overview of the data and we present results obtained both on the training data (estimating the generalization error) and on the test data (the actual prediction error). We also explore the use of category-specific decisions, as opposed to a global decision layer, as used for the competition. We also show the keywords extracted for each category and compare those to the official class description provided after the contest.

10.2 The Probabilistic Model

Let us first formalize the text categorization problem, such as proposed in the Anomaly Detection/Text Mining competition. We are provided with a set of M documents $\{d_i\}_{i=1...M}$ and associated labels $\ell \in \{1, ... C\}$, where C is the number of categories. These form the training set $\mathcal{D} = \{(d_i, \ell_i)\}_{i=1...M}$. Note that, for now, we will assume that there is only one label per document. We will address the multi-label situation later in Section 10.2.3. The text categorization task is the following: given a new document $\tilde{d} \notin \mathcal{D}$, find the most appropriate label $\tilde{\ell}$. There are mainly two types of inference for solving this problem [Vap98]. *Inductive* inference will estimate a model $\hat{f}$ using the training data $\mathcal{D}$, then assign $\tilde{d}$ to category $\hat{f}(\tilde{d})$. *Transductive* inference will estimate the label $\tilde{\ell}$ directly without estimating a general model.

We will see that our probabilistic model shares similarities with both. We estimate some model parameters, as described in Section 10.2.1, but we do not use the model directly to provide the label of new documents. Rather, prediction is done by estimating the labeling probabilities by maximizing the likelihood on the new document using an EM-type algorithm, cf. Section 10.2.2.

Let us now assume that each document d is composed of a number of words w from a vocabulary $\mathcal{V}$. We use the bag-of-words assumption. This means that the actual order of words is discarded and we use only the frequency $n(w, d)$ of each word w in each document d. The categorizer presented in [GGPC02] is a model of the *co-occurrences* (w, d). The probability of a co-occurrence, $P(w, d)$, is a mixture of C multinomial components, assuming one component per category:

$$P(w, d) = \sum_{c=1}^{C} P(c)P(d|c)P(w|c) \tag{10.1}$$

$$= P(d) \sum_{c=1}^{C} P(c|d)P(w|c) .$$

This is in fact the model used in *probabilistic latent semantic analysis* (PLSA) (cf. [Hof99]), but used in a supervised learning setting. The key modeling aspect is that documents and words are conditionally independent, which means that within each component, all documents use the same vocabulary in the same way. Parameters $P(w|c)$ are the profiles of each category (over the vocabulary), and parameters $P(c|d)$ are the profiles of each document (over the categories). We will now show how these parameters are estimated from the training data.

10.2.1 Training

The (log-)likelihood of the model with parameters $\theta = \{P(d); P(c|d); P(w|c)\}$ is

$$\mathcal{L}(\theta) = \log P(\mathcal{D}|\theta) = \sum_d \sum_{w \in \mathcal{V}} n(w, d) \log P(w, d), \qquad (10.2)$$

assuming independently identically distributed (iid) data.

Parameter estimation is carried out by maximizing the likelihood. Assuming that there is a one-to-one mapping between categories and components in the model, we have, for each training document, $P(c = \ell_i|d_i) = 1$ and $P(c \neq \ell_i|d_i) = 0$, for all i. This greatly simplifies the likelihood, which may now be maximized analytically. Let us introduce $|d| = \sum_w n(w, d)$ the length of document d, $|c| = \sum_{d \in c} |d|$ the total size of category c (using the shorthand notation $d \in c$ to mean all documents d_i such that $\ell_i = c$), and $N = \sum_d |d|$ the number of words in the collection. The maximum likelihood (ML) estimates are

$$\widehat{P}(w|c) = \frac{1}{|c|} \sum_{d \in c} n(w, d) \quad \text{and} \quad \widehat{P}(d) = \frac{|d|}{N}. \qquad (10.3)$$

Note that in fact only the category profiles $\widehat{P}(w|c)$ matter. As shown below, the document probability $\widehat{P}(d)$ is not used for categorizing new documents (as it is irrelevant, for a given d).

The ML estimates in Eq. (10.3) are essentially identical to those of the naive Bayes categorizer [MN98]. The underlying probabilistic models, however, are definitely different, as illustrated in Figure 10.1 and shown in the next section. One key consequence is that the probabilistic model in Eq. (10.1) is much less sensitive to smoothing than the naive Bayes.

It should be noted that the ML estimates rely on simple corpus statistics and can be computed in a single pass over the training data. This contrasts with many training algorithms that rely on iterative optimization methods. It means that training our model is extremely computationally efficient.

10.2.2 Prediction

Note that Eq. (10.1) is a generative model of co-occurrences of words and documents *within a given collection* $\{d_1 \ldots d_M\}$ *with a set vocabulary* $\mathcal{V}$. It is *not* a generative model of new documents, contrary to, for example, naive Bayes. This means that we cannot directly calculate the posterior probability $P(\widetilde{d}|c)$ for a new document.

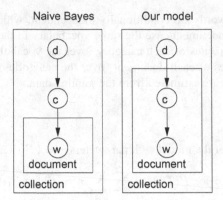

Fig. 10.1. Graphical models for naive Bayes (left) and for the probabilistic model used here (right).

We obtain predictions by *folding in* the new document in the collection. As document $\tilde{d}$ is folded in, the following parameters are added to the model: $P(\tilde{d})$ and $P(c|\tilde{d}), \forall c$. The latter are precisely the probabilities we are interested in for predicting the category labels. As before, we use a maximum likelihood (ML) approach, maximizing the likelihood for the new document:

$$\tilde{\mathcal{L}} = \sum_w n(w, \tilde{d}) \log P(\tilde{d}) \sum_c P(c|\tilde{d}) P(w|c) \tag{10.4}$$

with respect to the unknown parameters $P(c|\tilde{d})$.

The likelihood may be maximized using a variant of the expectation maximization (EM) (cf. [DLR77]) algorithm. It is similar to the EM used for estimating the PLSA model (see [Hof99, GGPC02]), with the constraint that the category profiles $P(w|c)$ are kept fixed. The iterative update is given by

$$P(c|\tilde{d}) \leftarrow P(c|\tilde{d}) \sum_w \frac{n(w, \tilde{d})}{|\tilde{d}|} \frac{P(w|c)}{\sum_c P(c|\tilde{d}) P(w|c)}. \tag{10.5}$$

The likelihood (Eq. (10.4)) is guaranteed to be strictly increasing with every EM step, therefore Eq. (10.5) converges to a (local) minimum. In the general case of unsupervised learning, the use of deterministic annealing [RGF90] during parameter estimation helps reduce sensitivity to initial conditions and improves convergence (cf. [Hof99, GGPC02]). Note, however, that as we only need to optimize over a small set of parameters, such annealing schemes are typically not necessary at the prediction stage. Upon convergence, the posterior probability estimate for $P(c|\tilde{d})$ may be used as a basis for assigning the final category label(s) to document $\tilde{d}$.

Readers familiar with nonnegative matrix factorization (NMF) (cf. [LS99]) will have noticed that Eq. (10.5) is very similar to one of the update rules used for estimating the NMF that minimizes a KL-type divergence between the data and the model.

Indeed, the unsupervised counterpart to our model, PLSA, is essentially equivalent to NMF in this context [GG05a]. Therefore, another way of looking at the categorization model described in this chapter is in fact to view it as a *constrained NMF* problem, with the category labels providing the constraint on one of the factors (the loading).

The way the prediction is obtained also sheds some light on the difference between our method and a naive Bayes categorizer. In naive Bayes, a category is associated with a whole document, and all words from this document must then be generated from this category. The occurrence of a word with a low probability in the category profile will therefore impose an overwhelming penalty to the category posterior $P(c|d)$. By contrast, the model we use here assigns a category c to each *co-occurrence* (w, d), which means that each *word* may be sampled from a different category profile. This difference manifests itself in the re-estimation formula for $P(c|\tilde{d})$, Eq. (10.5), which combines the various word probabilities as a *sum*. As a consequence, a very low probability word will have little influence on the posterior category probability and, more importantly, will not impose an overwhelming penalty. This key difference also makes our model much less sensitive to probability smoothing than naive Bayes. This means that we do not need to set extra parameters for the smoothing process. In fact, up to that point, we do not need to set *any* extra hyperparameter in either the training or the prediction phases.

As an aside, it is interesting to relate our method to the two paradigms of *inductive* and *transductive* learning [Vap98]. The *training* phase seems typically *inductive*: we optimize a cost function (the likelihood) to obtain one optimal model. Note, however, that this is mostly a model of the *training* data, and it does not provide direct labeling for any document outside the training set. At the *prediction* stage, we perform another optimization, this time over the labeling of the test document. This is in fact quite similar to *transductive* learning. As such, it appears that our probabilistic model shares similarities with both learning paradigms.

We will now address two important issues of the Anomaly Detection/Text Mining competition that require some extensions to the basic model that we have presented. Multilabel categorization is addressed in Section 10.2.3 and the estimation of a prediction confidence is covered in Section 10.2.4.

10.2.3 Multiclass, Multilabel Categorization

So far, the model we have presented is strictly a multiclass, *single-label* categorization model. It can handle more than two classes ($C > 2$), but the random variable c indexing the categories takes a *single* value in a discrete set of C possible categories.

The Anomaly Detection/Text Mining competition is a multiclass, *multilabel* categorization problem; that is, each document may belong to multiple categories. In fact, although most documents have only one or two labels, one document, number 4898, has 10 labels (out of 22), and five documents have exactly nine labels.

One principled way to extend our model to handle multiple labels per document is to consider all observed combinations of categories and use these combinations as single "labels," as described in [McC99]. On the competition data, however, there are 1151 different label combinations with at least one associated document. This makes this approach hardly practical. An additional issue is that considering label combinations independently, one may miss some dependencies between single categories. That is, one can expect that combinations $(C4, C5, C10)$ and $(C4, C5, C11)$ may be somewhat dependent as they share two out of three category labels. This is not modeled by the basic "all model combinations" approach. Although dependencies may be introduced as described, for example, in [GG06], this adds another layer of complexity to the system. In our case, the number of dependencies to consider between the 1151 observed label combinations is overwhelming.

Another approach is to reduce the multiple labeling problem to a number of binary categorization problems. With 22 possible labels, we would therefore train 22 binary categorizers and use them to make 22 independent labeling decisions. This is an appealing and usually successful approach, especially with powerful binary categorizers such as support vector machines [Joa98]. However, it still mostly ignores dependencies between the individual labels (for example, the fact that labels $C4$ and $C5$ are often observed together), and it multiplies the training effort by the number of labels (22 in our case).

Our approach is actually somewhat less principled than the alternatives mentioned above, but a lot more straightforward. We rely on a simple threshold $a \in [0; 1]$ and assign any new document $\tilde{d}$ to all categories c such that $P(c|\tilde{d}) \geq a$. In addition, as all documents in the training set have at least one label, we make sure that $\tilde{d}$ always gets assigned the label with the highest $P(c|\tilde{d})$, even if this maximum is below the threshold. This threshold is combined with the calculation of the confidence level, as explained in the next section.

10.2.4 Confidence Estimation

A second important issue in the Anomaly Detection/Text Mining competition is that labeling has to be provided with an associated confidence level.

The task of estimating the proper probability of correctness for the output of a categorizer is sometimes called *calibration* [ZE01]. The confidence level is then the probability that a given labeling will indeed be correct, that is, labels with a confidence of 0.8 will be correct 80% of the time. Unfortunately, there does not seem to be any guarantee that the cost function used for the competition will be optimized by a "well-calibrated" confidence (cf. Section 10.3.2 below). In fact, there is always a natural tension between calibration and performance. Some perfectly calibrated categorizers can show poor performance; conversely, some excellent categorizers (for example, support vector machines) may be poorly or not calibrated.

Accordingly, instead of seeking to calibrate the categorizer, we use the provided score function, Checker.jar, to optimize a function that outputs the confidence level, given the probability output by the categorizer. In fields like speech recognition, and more generally in natural language processing, confidence estimation is often done by adding an additional machine learning layer to the model [GIY97, GFL06], using the output of the model and possibly additional external features measuring the level of difficulty of the task. We adopt a similar approach, but use a much simpler model.

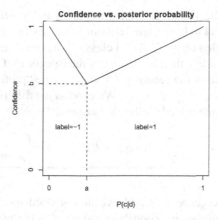

Fig. 10.2. The piecewise linear function used to transform the posterior probability into a confidence level.

The confidence layer transforms the conditional probability output by the model, $P(c|\tilde{d})$, into a proper confidence measure by using a piecewise linear function with two parameters (Figure 10.2). One parameter is the probability threshold a, which determines whether a label is assigned or not; the second is a baseline confidence level b, which determines what confidence we give a document that is around the threshold. The motivation for the piecewise-linear shape is that it seems reasonable that the confidence is a monotonic function of the probability; that is, if two documents $\tilde{d}_1$ and $\tilde{d}_2$ are such that $a < P(c|\tilde{d}_1) < P(c|\tilde{d}_2)$, then it makes sense to give $\tilde{d}_2$ a higher confidence to have label c than $\tilde{d}_1$. Using linear segments is a parsimonious way to implement this assumption.

Let us note that the entire model, including the confidence layer, relies on only two learning parameters, a and b. These parameters may be optimized by maximizing the score obtained on a prediction set or a cross-validation estimator, as explained below.

10.2.5 Category Description

The model relies on probabilistic profiles $P(w|c)$ that represent the probability of each word of the vocabulary to be observed in documents of category c. These profiles allow us to identify which words are more typical of each category, and therefore may provide a way to interpret the data by providing descriptive keywords for each category.

Notice that the simplest way of doing this, by focusing on words w with the highest probabilities $P(w|c)$, is not very efficient. First of all, the probability profile is linked to the frequency of words in the training corpus (Eq. (10.3)), and high-frequency words tend to be grammatical ("empty") words with no descriptive content. Even when grammatical words have been filtered out, words typical of the general topic of the collection (e.g., related to planes and aviation in the contest corpus, see below) will tend to have high frequency in many categories.

To identify words that are typical of a class c, we need to identify words that are *relatively* more frequent in c than in the rest of the categories. One way of doing that is to contrast the profile of the category $P(w|c)$ and the "profile" for the rest of the data, that is, $P(w|\neg c) \propto \sum_{\gamma \neq c} P(w|\gamma)$. We express the difference between the two distributions by the symmetrized Kullback-Leibler divergence:

$$KL(c, \neg c) = \sum_{w} \underbrace{(P(w|c) - P(w|\neg c)) \log \frac{P(w|c)}{P(w|\neg c)}}_{k_w} \,. \qquad (10.6)$$

Notice that the divergence is an additive sum of word-specific contributions k_w. Words with a large value of k_w contribute the most to the overall divergence, and hence to the difference between category c and the rest. As a consequence, we propose as keywords the words w for which $P(w|c) > P(w|\neg c)$ and k_w is the largest.[2] In the following section, we will see how this strategy allows us to extract keywords that are related to the actual description of the categories.

10.3 Experimental Results

We will now describe some of our experiments in more details and give some results obtained both for the estimated prediction performance, using only the training data provided for the competition, and on the test set using the labels provided after the competition.

10.3.1 Data

The available training data consists of 21,519 reports categorized in up to 22 categories. Some limited preprocessing was performed by the organizers, such as tokenization, stemming, acronym expansion, and removal of places and numbers. This preprocessing makes it nontrivial for participants to leverage their own in-house linguistic preprocessing. On the other hand, it places contestants on a level playing field, which puts the emphasis on differences in the actual categorization method, as opposed to differences in preprocessing.[3]

The only additional preprocessing we performed on the data was stop-word removal, using a list of 319 common words. Similar lists are available many places on the Internet. After stop-word removal, documents were indexed in a bag-of-words format by recording the frequency of each word in each document.

To obtain an estimator of the prediction error, we organized the data in a 10-fold cross-validation manner. We randomly reordered the data and formed 10 splits:

---

[2] Alternatively, we can rank words according to $\tilde{k}_w = k_w \text{sign}(P(w|c) - P(w|\neg c))$.

[3] In our experience, differences in preprocessing typically yield larger performance gaps than differences in categorization method.

nine containing 2152 documents, and one with 2151 documents. We then trained a categorizer using each subset of nine splits as training material, as described in Section 10.2.1, and produced predictions on the remaining split, as described in Section 10.2.2. As a result, we obtain 21,519 predictions on which we optimized parameters a and b.

Note that the 7077 test data on which we obtained the final results reported below were never used during the estimation of either model parameters or additional decision parameters (thresholds and confidence levels).

10.3.2 Results

The competition was judged using a specific cost function combining prediction performance and confidence reliability. For each category c, we compute the area under the ROC curve, A_c, for the categorizer. A_c lies between 0 and 1, and is usually above 0.5. In addition, for each category c, denote $t_{ic} \in \{-1, +1\}$ the target label for document d_i, $y_{ic} \in \{-1, +1\}$ the predicted label and $q_{ic} \in [0; 1]$ the associated confidence. The *final cost function* is

$$Q = \frac{1}{C} \sum_{c=1}^{C} (2A_c - 1) + \frac{1}{M} \sum_{i=1}^{M} q_{ic} t_{ic} y_{ic} . \tag{10.7}$$

Given predicted labels and associated confidence, the reference script Checker.jar provided by the organizers computes this final score. A perfect prediction with 100% confidence yields a final score of 2, while a random assignment would give a final score of around 0.

Using the script Checker.jar on the cross-validated predictions, we optimized a and b using alternating optimizations along both parameters. The optimal values used for our submission to the competition are $a = 0.24$ and $b = 0.93$, indicating that documents are labeled with all categories that have a posterior probability higher than 0.24, and the minimum confidence is 0.93. This seems like an unusually high baseline confidence (i.e., we label with *at least* 93% confidence). However, this is not surprising if one considers the expression of the cost function (Eq. (10.7)) more closely. The first part is the area under the curve, which depends only on the ordering. Although the ordering is based on the confidence levels, it only depends on the relative, not the absolute, values. For example, with our confidence layer, the ordering is preserved regardless of the value of b, up to numerical precision.

On the other hand, the second part of the cost function (Eq. (10.7)) directly involves the confidence estimates q_{ic}. The value of this part will increase if we can reliably assign high confidence ($q_{ic} \approx 1$) to correctly labeled documents ($t_{ic} = y_{ic}$) and low confidence ($q_{ic} \approx 0$) to incorrectly labeled documents ($t_{ic} \neq y_{ic}$). However, if we could reliably detect such situations, we would arguably be better off swapping the label rather than downplay its influence by assigning it low confidence. So assuming we cannot reliably do so, that is, q_{ic} and $t_{ic} y_{ic}$ are independent, the second part of the cost becomes approximately equal to $\bar{q}(2 \times MCE - 1)$, with $\bar{q}$ the average confidence and MCE the *misclassification error*, $MCE = 1/M \sum_i (t_{ic} \neq y_{ic})$. So

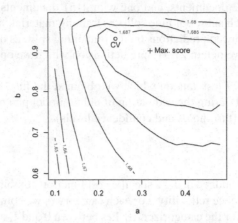

Fig. 10.3. Score for various combinations of a and b. The best (maximum) test score is indicated as a cross, the optimum estimated by cross-validation (CV) is $a = 0.24$ and $b = 0.93$, indicated by a circle.

the optimal strategy under this assumption is to make $\bar{q}$ as high as possible by setting q_{ic} as close to 1 as possible, while keeping the ordering intact. This explains why a relatively high value of b turned out to be optimal. In fact, using a higher precision in our confidence levels, setting b to 0.99 or higher yields even better results.[4]

Using the setting $a = 0.24$ and $b = 0.93$, the cross-validated cost is about 1.691. With the same settings, the final cost on the 7077 test documents is 1.689, showing an excellent agreement with the cross-validation estimate. To illustrate the sensitivity of the performance to the setting of the two hyperparameters a and b, we plot the final cost obtained for various combinations of a and b, as shown in Figure 10.3. The optimal setting (cross) is in fact quite close to the cross-validation estimate (circle). In addition, it seems that the performance of the system is not very sensitive to the precise values of a and b. Over the range plotted in Figure 10.3, the maximal score (cross) is 1.6894, less than 0.05% above the CV-optimized value, and the lowest score (bottom left) is 1.645, 2.5% below. This means that any setting of a and b in that range would have been within 2.5% of the optimum.

We also measured the performance using some more common and intuitive metrics. For example, the overall mislabeling error rate is 7.22%, and the micro-averaged F-score is a relatively modest 50.05%.

Note, however, that we have used a single confidence layer, and in particular a single labeling threshold a, for all categories. Closer inspection of the performance on each category shows quite a disparity in performance, and in particular in pre-

---

[4] Note, however, that setting b to higher value only yields marginal benefits in terms of final cost.

Table 10.1. Performance of the probabilistic model: precision, recall, and F-score for each of the 22 categories. Low ($< 50\%$) scores are in italics and high ($> 75\%$) scores are in bold. Column "# docs" contains the number of test documents with the corresponding label.

Category	#docs	p (%)	r (%)	F (%)	Category	#docs	p (%)	r (%)	F (%)
C1	435	64.99	**83.22**	72.98	C12	918	72.02	**79.08**	**75.39**
C2	3297	*46.72*	**99.57**	63.60	C13	687	51.21	55.46	53.25
C3	222	74.06	**79.73**	**76.79**	C14	393	68.73	70.48	69.60
C4	182	52.15	59.89	55.75	C15	183	*30.61*	*24.59*	*27.27*
C5	853	**80.07**	**75.38**	**77.66**	C16	314	*32.93*	60.83	*42.73*
C6	1598	51.33	**80.98**	62.83	C17	162	*45.22*	*43.83*	*44.51*
C7	546	*36.38*	**77.29**	*49.47*	C18	351	57.30	58.12	57.71
C8	631	55.32	67.51	60.81	C19	1767	61.54	**80.42**	69.73
C9	168	59.30	60.71	60.00	C20	229	72.40	69.87	71.11
C10	349	*37.21*	58.74	*45.56*	C21	137	**78.79**	56.93	66.10
C11	161	**77.44**	63.98	70.06	C22	233	**88.66**	73.81	**80.56**

cision and recall, across the categories.[5] This suggests that the common value of the threshold a may be too high for some categories (hence low recall) and too low for others (hence low precision). Using the cross-validated prediction, we therefore optimized some category-specific thresholds using various metrics:

- Maximum F-score
- Break-even point (i.e., point at which precision equals recall)
- Minimum error rate

For example, for maximum F-score, we optimized 22 thresholds, one per category, by maximizing the F-score for each category on the cross-validated predictions.

Table 10.1 shows the performance obtained on the test data for all 22 categories, using category-specific, maximum F-score optimized thresholds. Performance is expressed in terms of the standard metrics of *precision*, *recall*, and *F-score*. The performance varies a lot depending on the category. However, there does not seem to be any systematic relation between the performance and the size of the categories. The worst, but also the best performance are observed on small categories (less that 250 positive test documents). This suggests that the variation in performance may be simply due to varying intrinsic difficulty of modeling the categories. The best F-scores are observed on categories C3, C5, C12 and C22, while categories C15, C16, C17, C10 and C7 get sub-50% performance.

The average performance is presented in Table 10.2 for our submission to the competition, as well as for three different strategies for optimizing a category-per-category threshold. One weakness of the original submission was the low recall, due to the fact that a single threshold was used for all 22 categories. This produces a relatively low average F-score of 50.05%. By optimizing the threshold for each

---

[5] *Precision* estimates the probability that a label provided by the model is correct, while *recall* estimates the probability that a reference label is indeed returned by the model [GG05b]. *F-score* is the harmonic average of precision and recall.

Table 10.2. Micro-averaged performance for our contest submission, and three threshold optimization strategy. Highest is best for precision (p), recall (r), F-score (F), and final cost (Eq. (10.7)), and lowest is best for misclassification error (MCE).

	p (%)	r (%)	F (%)	MCE (%)	Final cost
Our submission	**64.87**	*40.74*	*50.05*	7.22	1.689
Maximum F-score	*53.30*	**78.55**	**63.51**	*8.01*	*1.678*
Break-even point	58.88	67.35	62.83	7.07	1.697
Minimum error	61.53	62.37	61.95	**6.80**	**1.702**

category over various performance measures, we largely improve over that. Not surprisingly, optimizing the thresholds for F-scores yields the best F-score of 63.51%, although both the misclassification error and the final cost *degrade* using this strategy. Optimizing the threshold for reducing the misclassification error reduces the test misclassification error[6] to 6.80% and improves the final cost slightly, to 1.702. Note, however, that despite a large impact on F-score and MCE, using category-specific decision thresholds optimized over various thresholds seems to have little impact on the final cost, which stays within about 0.8% of the score of our submission.

10.3.3 Category Description

The only information provided by the organizers at the time of the contest was the labeling for each document, as a number between 1 and 22. These data, therefore, seemed like a good test bed for applying the category description technique described above, and to see whether the extracted keywords brought any information on the content of the categories.

Table 10.3 shows the results we obtained on about half the categories by extracting the top five keywords. For comparison, we also extracted the five words with highest probability (first column), and we also give the official category description from the Distributed National Aviation Safety Action Program Archive (DNAA), which was released after the competition.

Based on only five keywords, the relevance of the keywords provided by either highest probability or largest divergence and their mapping to the official category descriptions are certainly open to interpretation. The most obvious problem with the choice of the highest probability words, however, is that the same handful of keywords seem to appear in almost all categories. The words "aircraft" is among the top-five probability in 19 out of 22 categories! Although it is obviously a very relevant and common word for describing issues dealing with aviation, it is clearly not very useful to discriminate the content of one category versus the others. The other frequent members of the top-five highest probability are: runway (14 times), airport (11), approach (11), feet (9), and land (9). These tend to "pollute" the keyword extraction. For example, in category C8 (course deviation), three out of the five highest probability keywords are among these frequent keywords and bring no relevant

---

[6] For 7077 test documents with 22 possible labels, reducing the MCE by 0.1% corresponds to correcting 156 labeling decisions.

Table 10.3. Keywords extracted for 12 of the contest categories using either the highest probability words or the largest contributions to the KL divergence; for comparison, the official description from the DNAA, released after the competition, is provided.

Category	Highest probability	Largest divergence	Official DNAA category description
C1	aircraft maintain minimumequip-mentlist check flight	minimumequip-mentlist maintain install defer inspect	*Airworthiness/documentation event:* An event involving a incorrect or incomplete airworthi-ness or documentation requirement
C2	aircraft runway airport feet approach	security flight board agent airspace	*Operation in noncompliance—FARs, pol-icy/procedures:* An event involving a viola-tion, deviation, or noncompliance by the crew which involved a policy, procedure, or FAR
C3	runway takeoff aircraft clear tower	takeoff abort runway reject roll	*Rejected takeoff:* An event involving a re-jected takeoff
C4	runway aircraft land taxiway left	brake taxiway runway damage grass	*Excursion:* An event involving the loss of con-trol or inadvertent control of an aircraft from the designated airport surface
C5	runway aircraft taxiway taxi hold	runway taxiway taxi hold short	*Incursion:* An event involving a vehicle, per-son, object or other aircraft that creates a col-lision hazard or results in loss of separation with an aircraft
C8	runway turn airport degree approach	degree turn head radial course	*Course deviation:* An event involving a devia-tion from an assigned or planned course
C9	feet knot aircraft approach speed	knot speed slow knotsindi-catedairspeed airspeed	*Speed deviation:* An event involving a devia-tion from a planned or assigned air speed
C10	aircraft runway approach land feet	brake knot wind autopilot damage	*Uncommanded/unintended state or loss of control:* An event involving an uncommanded, unintended state or loss of control of the air-craft
C11	approach feet airport runway descend	approach groundproximi-tywarningsys-tem terrain feet glideslope	*Terrain proximity event:* An event involving the aircraft operating in close proximity to ter-rain
C13	airport feet approach aircraft runway	weather turbulent cloud encounter ice thunderstorm	*Weather issue/ weather proximity event:* An event involving a weather issue or aircraft op-erating in proximity to weather

descriptive information. The remaining two keywords, "turn" and "degree," on the other hand, seem like reasonable keywords to describe course deviation problems.

By contrast, the five words contributing to the largest divergence, "degree," "turn," "head," "radial," and "course," appear as keywords only for this category, and they all seem topically related to the corresponding problem. For the largest divergence metric, k_w, the word selected most often in the top-five is "runway," which describes six categories containing problems related to takeoff or landing (e.g., C3, C4, C5, and C6). Other top-five keywords appear in at most four categories. This suggests that using k_w instead of $P(w|c)$ yields more diverse and more descriptive keywords.

This is also supported by the fact that only 32 distinct words are used in the top-five highest probability keywords over the 22 categories, while the top-five largest k_w selects 83 distinct words over the 22 categories. This further reinforces the fact that k_w will select more specific keywords, and discard the words that are common to all categories in the corpus.

This is well illustrated on category C13 (Weather Issue). The top-five probability words are all among the high probability words mentioned above. By contrast, the top-5 k_w are all clearly related to the content of category C13, such as "weather," "cloud," "ice," and "thunderstorm."

Overall, we think that this example illustrates the shortcomings of the choice of the highest probability words to describe a category. It supports our proposition to use instead words that contribute most to the divergence between one category and the rest. These provide a larger array of keywords and seem more closely related to the specificities of the categories they describe.

10.4 Summary

We have presented the probabilistic model that we used in NRC's submission to the Anomaly Detection/Text Mining competition at the Text Mining Workshop 2007. This probabilistic model may be estimated from preprocessed, indexed, and labeled documents with no additional learning parameters, and in a single pass over the data. This makes it extremely fast to train. On the competition data, in particular, the training phase takes only a few seconds on a current laptop. Obtaining predictions for new test documents requires a bit more calculations but is still quite fast. One particularly attractive feature of this model is that the probabilistic category profiles can be used to provide descriptive keywords for the categories. This is useful in the common situation where labels are known only as codes and the actual content of the categories may not be known to the practitioner.

The only training parameters we used are required for tuning the decision layer, which selects the multiple labels associated to each document, and estimates the confidence in the labeling. In the method that we implemented for the competition, these parameters are the labeling threshold and the confidence baseline. They are estimated by maximizing the cross-validated cost function.

Performance on the test set yields a final score of 1.689, which is very close to the cross-validation estimate. This suggests that despite its apparent simplicity, the probabilistic model provides a very efficient categorization. This is actually corroborated by extensive evidence on multiple real-life use cases.

The simplicity of the implemented method, and in particular the somewhat rudimentary confidence layer, suggests that there may be ample room for improving the performance. One obvious issue is that the *ad-hoc* layer used for labeling and estimating the confidence may be greatly improved by using a more principled approach. One possibility would be to train multiple categorizers, both binary and multicategory. Then use the output of these categorizers as input to a more complex model combining this information into a proper decision associated with a better confidence level. This may be done, for example, using a simple logistic regression. Note that one issue here is that the final score used for the competition, Eq. (10.7), combines a performance-oriented measure (area under the ROC curve) and a confidence-oriented measure. As a consequence, and as discussed above, there is no guarantee that a well-calibrated classifier will in fact optimize this score. Also, this suggests that there may be a way to invoke multiobjective optimization in order to further improve the performance.

Among other interesting topics, let us mention the sensitivity of the method to various experimental conditions. In particular, although we have argued that our probabilistic model is not very sensitive to smoothing, it may very well be that a properly chosen smoothing, or similarly, a smart feature selection process, may further improve the performance. In the context of multilabel categorization, let us also mention the possibility to exploit dependencies between the classes, for example, using an extension of the method described in [RGG⁺06].

References

[DLR77]   A.P. Dempster, N.M. Laird, and D.B. Rubin. Maximum likelihood from incomplete data via the EM algorithm. *Journal of the Royal Statistical Society, Series B*, 39(1):1–38, 1977.

[GFL06]   S. Gandrabur, G. Foster, and G. Lapalme. Confidence estimation for NLP applications. *ACM Transactions on Speech and Language Processing,*, 3(3):1–29, 2006.

[GG05a]   E. Gaussier and C. Goutte. Relation between PLSA and NMF and implications. In *Proceedings of the 28th Annual International ACM SIGIR Conference*, pages 601–602, ACM Press, New York, 2005.

[GG05b]   C. Goutte and E. Gaussier. A probabilistic interpretation of precision, recall and F-score, with implication for evaluation. In *Advances in Information Retrieval, 27th European Conference on IR Research (ECIR 2005)*, pages 345–359. Springer, New York, 2005.

[GG06]   C. Goutte and E. Gaussier. Method for multi-class, multi-label categorization using probabilistic hierarchical modeling. US Patent 7,139,754, granted November 21, 2006.

[GGPC02] E. Gaussier, C. Goutte, K. Popat, and F. Chen. A hierarchical model for clustering and categorising documents. In *Advances in Information Retrieval*, Lecture Notes in Computer Science, pages 229–247. Springer, New York, 2002.

[GIY97]   L. Gillick, Y. Ito, and J. Young. A probabilistic approach to confidence estimation and evaluation. In *ICASSP '97: Proceedings of the 1997 IEEE International Conference on Acoustics, Speech, and Signal Processing*, volume 2, pages 879–882, IEEE, Los Alamitos, CA, 1997.

[Hof99]   T. Hofmann. Probabilistic latent semantic analysis. In K.B. Laskey and H. Prade, editors, *UAI '99: Proceedings of the Fifteenth Conference on Uncertainty in Artificial Intelligence*, pages 289–296. Morgan Kaufmann, San Francisco, 1999.

[Joa98]   T. Joachims. Text categorization with suport vector machines: learning with many relevant features. In Claire Nédellec and Céline Rouveirol, editors, *Proceedings of ECML-98, 10th European Conference on Machine Learning*, volume 1398 of *Lecture Notes in Computer Science*, pages 137–142. Springer, New York, 1998. Available from World Wide Web: `citeseer.ist.psu.edu/joachims97text.html`.

[LS99]   D. Lee and H. Seung. Learning the parts of objects by non-negative matrix factorization. *Nature*, 401:788–791, 1999.

[McC99]   A. McCallum. Multi-label text classification with a mixture model trained by EM. In *AAAI'99 Workshop on Text Learning*, 1999.

[MN98]   A. McCallum and K. Nigam. A comparison of event models for naive Bayes text classification. In *AAAI/ICML-98 Workshop on Learning for Text Categorization*, pages 41–48, AAAI Press, Menlo Park, CA, 1998.

[RGF90]   K. Rose, E. Gurewitz, and G. Fox. A deterministic annealing approach to clustering. *Pattern Recognition Letters*, 11(11):589–594, 1990.

[RGG$^+$06] J.-M. Renders, E. Gaussier, C. Goutte, F. Pacull, and G. Csurka. Categorization in multiple category systems. In *Machine Learning, Proceedings of the Twenty-Third International Conference (ICML 2006)*, pages 745–752, ACM Press, New York, 2006.

[Vap98]   V.N. Vapnik. *Statistical Learning Theory*. Wiley, New York, 1998.

[ZE01]   B. Zadrozny and C. Elkan. Obtaining calibrated probability estimates from decision trees and naive bayesian classifiers. In *Proceedings of the Eighteenth International Conference on Machine Learning (ICML 2001)*, pages 609–616, Morgan Kaufmann, San Francisco, 2001.

11

Anomaly Detection Using Nonnegative Matrix Factorization

Edward G. Allan, Michael R. Horvath, Christopher V. Kopek, Brian T. Lamb, Thomas S. Whaples, and Michael W. Berry

Overview

For the Text Mining 2007 Workshop contest (see Appendix), we use the nonnegative matrix factorization (NMF) to generate feature vectors that can be used to cluster Aviation Safety Reporting System (ASRS) documents. By preserving non-negativity, the NMF facilitates a sum-of-parts representation of the underlying term usage patterns in the ASRS document collection. Both the training and test sets of ASRS documents are parsed and then factored by the NMF to produce a reduced-rank representation of the entire document space. The resulting *feature* and *coefficient* matrix factors are used to cluster ASRS documents so that the (known) anomalies of training documents are directly mapped to the feature vectors. Dominant features of test documents are then used to generate anomaly relevance scores for those documents. The General Text Parser (GTP) software environment is used to generate term-by-document matrices for the NMF model.

11.1 Introduction

Nonnegative matrix factorization (NMF) has been widely used to approximate high-dimensional data comprised of nonnegative components. Lee and Seung [LS99] proposed the idea of using NMF techniques to generate basis functions for image data that could facilitate the identification and classification of objects. They also demonstrated the use of NMF to extract concepts/topics from unstructured text documents. This is the context that we exploit the so-called *sum-of-parts* representation offered by the NMF for the Aviation Safety Reporting System (ASRS) document collection.

Several manuscripts have cited [LS99], but as pointed out in [BBL+07] there are several (earlier) papers by P. Paatero [Paa97, Paa99, PT94] that documented the historical development of the NMF. Simply stated, the problem defining the NMF is as follows:

Given a nonnegative matrix $\mathbf{A} \in \mathbf{R}^{m \times n}$ and a positive integer $k < \min\{m, n\}$, find nonnegative matrices $\mathbf{W} \in \mathbf{R}^{m \times k}$ and $\mathbf{H} \in \mathbf{R}^{k \times n}$ to minimize the functional

$$f(\mathbf{W}, \mathbf{H}) = \frac{1}{2} \|\mathbf{A} - \mathbf{W}\mathbf{H}\|_{\mathbf{F}}^2. \tag{11.1}$$

The product $\mathbf{W}\mathbf{H}$ is called a nonnegative matrix factorization of $\mathbf{A}$, although $\mathbf{A}$ is not necessarily equal to the product $\mathbf{W}\mathbf{H}$. Although the product $\mathbf{W}\mathbf{H}$ is an approximate factorization of rank at most k, we will drop the word *approximate* in our NMF discussions below. The best choice for the rank k is certainly problem dependent, and in most cases it is usually chosen such that $k \ll \min(m, n)$. Hence, the product $\mathbf{W}\mathbf{H}$ can be considered a *compressed* form of the data in $\mathbf{A}$.

Another key characteristic of NMF is the ability of numerical methods that minimize Eq. (11.1) to extract underlying features as basis vectors in $\mathbf{W}$, which can then be subsequently used for identification and classification. By not allowing negative entries in $\mathbf{W}$ and $\mathbf{H}$, NMF enables a non-subtractive combination of parts to form a whole [LS99]. Features may be parts of faces in image data, topics or clusters in textual data, or specific absorption characteristics in hyperspectral data. In this chapter, we discuss the enhancement of NMF algorithms for the primary goal of feature extraction and identification in text and spectral data mining.

Important challenges affecting the numerical minimization of Eq. (11.1) include the existence of local minima due to the nonconvexity of $f(\mathbf{W}, \mathbf{H})$ in both $\mathbf{W}$ and $\mathbf{H}$. The nonuniqueness of its solution is easily realized by noting that $\mathbf{W}\mathbf{D}\mathbf{D}^{-1}\mathbf{H}$ for any nonnegative invertible matrix $\mathbf{D}$ whose inverse, $\mathbf{D}^{-1}$, is also nonnegative. Fortunately, the NMF is still quite useful for text/data mining in practice since even local minima can provide desirable data compression and feature extraction as will be demonstrated in this contest entry.

Alternative formulations of the NMF problem certainly arise in the literature. As surveyed in [BBL+07], an information theoretic formulation in [LS01] is based on the Kullback–Leibler divergence of $\mathbf{A}$ from $\mathbf{W}\mathbf{H}$, and the cost functions proposed in [CZA06] are based on Csiszár's φ-divergence. A formulation in [WJHT04] enforces constraints based on the Fisher linear discriminant analysis, and [GBV01] suggests using a diagonal weight matrix $\mathbf{Q}$ in the factorization model, $\mathbf{A}\mathbf{Q} \approx \mathbf{W}\mathbf{H}\mathbf{Q}$, as an attempt to compensate for feature redundancy[1] in the columns of $\mathbf{W}$. For other approaches using alternative cost functions, see [HB06] and [DS05].

To speed up convergence of Lee and Seung's (standard) NMF iteration, various alternative minimization strategies for Eq. (11.1) have been suggested. For example, [Lin05b] proposes the use of a projected gradient bound-constrained optimization method that presumably has better convergence properties than the standard multiplicative update rule approach. However, the use of certain auxiliary constraints in Eq. (11.1) may break down the bound-constrained optimization assumption and thereby limit the use of projected gradient methods. Accelerating the standard approach via an interior-point gradient method has been suggested in [GZ05], and a quasi-Newton optimization approach for updating $\mathbf{W}$ and $\mathbf{H}$, where negative values are replaced with small positive ϵ parameter to enforce nonnegativity, is discussed in

---

[1] Such redundancy can also be alleviated by using column stochastic constraints on $\mathbf{H}$ [PPA07].

[ZC06]. For a more complete overview of enhancements to improve the convergence of the (standard) NMF algorithm, see [BBL+07].

Typically, $\mathbf{W}$ and $\mathbf{H}$ are initialized with random nonnegative values to start the standard NMF algorithm. Another area of NMF-related research has focused on alternate approaches for initializing or seeding the algorithm. The goal, of course, is to speed up convergence. In [WCD03] spherical k-means clustering is used to initialize $\mathbf{W}$ and in [BG05] singular vectors of $\mathbf{A}$ are used for initialization and subsequent cost function reduction. Optimal initialization, however, remains an open research problem.

11.2 NMF Algorithm

As surveyed in [BBL+07], there are three general classes of NMF algorithms: multiplicative update algorithms, gradient descent algorithms, and alternating least squares algorithms. For this study, we deployed the most basic multiplicative update method (initially described in [LS01]). This approach, based on a mean squared error objective function, is illustrated below using MATLAB$^{\circledR}$ array operator notation:

Algorithm 11.2.1 Multiplicative Update Algorithm for NMF

$\mathbf{W}$ = rand(m,k);   % $\mathbf{W}$ initially random
$\mathbf{H}$ = rand(k,n);   % $\mathbf{H}$ initially random
for i = 1 : maxiter
       $\mathbf{H} = \mathbf{H}$.* $(\mathbf{W}^{\mathrm{T}}\mathbf{A})$ ./ $(\mathbf{W}^{\mathrm{T}}\mathbf{W}\mathbf{H} + \epsilon)$;
       $\mathbf{W} = \mathbf{W}$.* $(\mathbf{A}\mathbf{H}^{\mathrm{T}})$ ./ $(\mathbf{W}\mathbf{H}\mathbf{H}^{\mathrm{T}} + \epsilon)$;
end

The parameter $\epsilon = 10^{-9}$ is added to avoid division by zero. As explained in [BBL+07], if this multiplicative update NMF algorithm converges to a stationary point, there is no guarantee that the stationary point is a local minimum for the objective function. Additionally, if the limit point to which the algorithm has converged lies on the boundary of the feasible region, we cannot conclude that it is, in fact, a stationary point. A modification of the Lee and Seung multiplicative update scheme that resolves some of the convergence issues and guarantees convergence to a stationary point is provided in [Lin05a].

11.3 Document Parsing and Term Weighting

The General Text Parsing (GTP) software environment [GWB03] (written in C++) was used to parse all the ASRS training and test documents for this contest. Let $\mathbf{A} = [\mathbf{R}|\mathbf{T}] = [a_{ij}]$ define an $m \times n$ term-by-document matrix, where the submatrices $\mathbf{R}$ and $\mathbf{T}$ represent training and test documents, respectively. Each element or component a_{ij} of the matrix A defines a *weighted* frequency at which term i occurs in document j. We can define $a_{ij} = l_{ij}g_i d_j$, where l_{ij} is the local weight for term i occurring in document j, g_i is the global weight for term i in the subcollection, and d_j is a document normalization factor that specifies whether or not the columns of A (i.e., the documents) are normalized (i.e., have unit length). Let f_{ij} be the number of times (frequency) that term i appears in document j, and define $\hat{p}_{ij} = f_{ij}/\sum_j f_{ij}$, that is, the empirical probability of term i appearing in document j. Using GTP, we deploy a log-entropy term-weighting scheme whereby

$$l_{ij} = \log(1 + f_{ij}) \text{ and } g_i = 1 + \left(\sum_j \hat{p}_{ij} log(\hat{p}_{ij})\right)/\log n \ .$$

No document normalization was used in parsing ASRS documents, that is, $d_j = 1$ for all a_{ij}. By default, GTP requires that the global frequency of any term, that is, $\sum_{j=1}^{n} f_{ij}$, be greater than 1 and that a term's document frequency (or number of documents containing that term) be greater than 1 as well. No adjustments to these thresholds were made in parsing the ASRS documents. A *stoplist* of 493 words[2] was used by GTP to filter out unimportant terms. The minimum and maximum length (in characters) of any accepted term was 2 and 200, respectively. For the initial training set of ASRS documents, GTP extracted $m = 15,722$ terms from the $n = 21,519$ documents. The elapsed GTP parsing time was 21.335 CPU seconds on a 3.2-GHz Intel Xeon having a 1024-KB cache and 4.1-GB RAM.

11.4 NMF Classifier

The classification of ASRS documents using NMF is outlined in Tables 11.1 and 11.2. Let H_i denote the ith column of matrix H. Some of the constants used include

- α, the threshold on the relevance score or (target value) t_{ij} for document i and anomaly/label j;

- δ, the threshold on the column elements of H, which will filter out the association of features with both the training (R) and test (T) documents;

- σ, the percentage of documents used to define the training set (or number of columns of R).

11.4.1 Preliminary Testing

The rank or number of columns of the feature matrix factor W used to test our NMF model (prior to contest submission) was $k = 40$. Hence, the W and H matrix factors were $15,722 \times 40$ and $40 \times 21,519$, respectively. The percentage of ASRS documents used for training (subset R) in our testing was 70% (i.e., $\sigma = .70$). Hence, 15,063 documents were used as the initial training set (R) and 6,456 documents were used

---

[2] See SMART's English stoplist at ftp://ftp.cs.cornell.edu/pub/smart/english.stop.

for testing (**T**) our NMF classifier. In step 1 of Table 11.1 we chose $\delta = .30$ for the columnwise pruning of the elements in the coefficient matrix **H**. This parameter effectively determines the number of features (among the $k = 40$ possible) that any document (training or test) can be associated with. As δ increases, so does the sparsity of **H**.

The α parameter specified in step 8 of Tables 11.1 and 11.2 was defined to be .40. This is the prediction control parameter that ultimately determines whether or not document i will be given label (anomaly) j, that is, whether $p_{ij} = +1$ or $p_{ij} = -1$ for the contest cost function

$$Q = \frac{1}{C} \sum_{j=1}^{C} Q_j, \tag{11.2}$$

$$Q_j = (2A_j - 1) + \frac{1}{D} \sum_{i=1}^{D} q_{ij} t_{ij} p_{ij}, \tag{11.3}$$

where C is the number of labels (anomalies) and D is the number of test documents. As mentioned above, $D = 6{,}456$ for our preliminary evaluation of the NMF classifier and $C = 22$ (as specified by the TRAINCATEGORYMATRIX.CSV file). The cost Q given by Eq. 11.2 for our preliminary NMF testing was always in the interval $[1.28, 1.30]$. To measure the quality of (anomaly) predictions across all $C = 22$ categories, a figure of merit (FOM) score defined by

$$FOM = \frac{1}{C} \sum_{j=1}^{C} \frac{F - F_j}{F} Q_j, \; F = \sum_{j=1}^{C} F_j, \tag{11.4}$$

where F_j denotes the frequency of documents having label (anomaly) j, and was generated for each experiment. By definition, the FOM score will assign lower weights to the higher frequency labels or categories. The best FOM score for $\sigma = .70$ was 1.267 to three significant decimal digits. Keep in mind that the initial matrix factors **W** and **H** are randomly generated and will produce slightly different features (columns of **W**) and coefficients (columns of **H**) per NMF iteration.[3]

11.4.2 Contest Results

For the text mining contest (sponsored by NASA Ames Research Center) at the Seventh SIAM International Conference on Data Mining in Minneapolis, MN (April 26–28, 2007), all contestants were provided an additional 7,077 ASRS unclassified documents. For the NMF classifier, we treat these *new* documents as the test subset **T** and define the training subset **R** as all the previously available (classified) documents (21,519 of them). Since all of the previously classified ASRS documents were used in the term-by-document matrix **A** for the contest entry, the σ parameter was set to 1.0. The other two parameters for the NMF classifier were not changed, that

---

[3] Only five iterations were used in our preliminary study.

Table 11.1. NMF-based classifier for ASRS documents with corresponding MATLAB® functions: steps 1–6

Step	Description	MATLAB® function call
1	Filter elements of **H** given **A** ≈ **WH**; for $i = 1, \ldots, n$, determine $\eta_i = \max(\mathbf{H}_i)$ and zero out all values in $\mathbf{H}_i$ less than $\eta_i \times (1 - \delta)$.	hThresh = asrsThreshold(H,DELTA);
2	Normalize the (new) filtered matrix **H** so that all column sums are now 1.	hNorm = asrsNorm(hThresh,K);
3	Given the released set of ASRS documents, generate a set of indices (integers) that will partition the documents into the training (**R**) and test (**T**) subsets.	[trainBinarySet, testBinarySet] = asrsGenerateSet(N,SIGMA);
4	Generate the corresponding submatrices of the normalized **H** matrix that correspond to **R** (hTrain) and **T** (hTest); generate submatrices for the anomaly matrix (csv file) so that anTrain and anTest are arrays for the document subsets **R** and **T**, respectively.	[hTrain hTest anTrain anTest] = asrsMapSets(hNorm,AN, trainBinarySet);
5	Cluster the columns of **H** corresponding to documents in the training set **R** by known anomalies (labels).	hClus = asrsClus(hTrain,K);
6	Sum the number of documents associated with each anomaly per NMF feature (k of them); the (i, j) entry of the output array hClusAnom specifies the number of anomaly/label j documents represented by feature i.	hClusAnom = asrsAnomClus(hClus,anTrain,K);

Table 11.2. NMF-based classifier for ASRS documents with corresponding MATLAB® functions: steps 7–10

Step	Description	MATLAB® function call
7	For each document in subset **T**, produce a score (or probability) that the document is relevant to each anomaly; loop over the number of NMF features (k of them) using the columns of the hTest and hClus arrays and use entries of hTest as *weights* on the relative frequency of any anomaly found in the current feature.	```hResults = asrsTesting(hTest,hClus,``` ```hClusAnom, K);```
8	Using α, produce the relevance/target score t_{ij} for (document i, anomaly j) pairs using the array hResults; the score will yield a positive prediction if $t_{ij} > \rho_i \times (1 - \alpha)$, where $\rho_i = max(\mathbf{H_i^T})$, i.e., the maximum element of the row of hResults corresponding to document i; entries of the output array prediction are either $+1$ or -1 to reflect positive or negative predictions, respectively; the confidence array contains the associated target values t_{ij}.	```[prediction,confidence] =``` ```asrsComputePC(hResults,ALPHA);```
9	Produce a component-wise matrix of predictions (p_{ij}) times confidences (t_{ij}) in the output array answers that can be visualized; create output arrays of correct (testResults) and incorrect (wrongAnswers) answers whereby rows correspond to documents and columns correspond to anomalies (or labels).	```[answers, testResults,``` ```wrongAnswers] = asrsComputeAnR``` ```(prediction,confidence,anTest);```
10	Generate ROC curves (and areas underneath) per anomaly and evaluate the contest cost function, and figure of merit score	```for i = 1:N``` ```[tp,fp] = roc(anTest(:,i),``` ```hResults(:,i));``` ```rocarea(i)=auroc(tp,fp);``` ```end;``` ```[cost,fom] = asrsCost(anTest,``` ```prediction,confidence,rocarea');```

is, $\alpha = 0.40$ and $\delta = 0.30$ (see Section 11.4.1). Using five iterations and $k = 40$ features for the multiplicative update algorithm mentioned in Section 11.2, a cost of $Q = 1.27$ (see Eq. 11.2) was reported by contest officials for the NMF classifier in mapping each of the 7,077 test documents to any of the 22 anomaly categories was 1.27 (a second place finish). Had a tie occurred among any of the cost function values generated by contest entries, the FOM score would have been used to break it. For the NMF classifier, the average contest FOM score was 1.22 (slightly lower than what was observed in the preliminary testing phase).

11.4.3 ROC Curves

Figures 11.1 through 11.5 contain the receiver operating characteristic (ROC) curves for the NMF classifier, when applied to selected training documents in the preliminary testing phase and to the unclassified test documents in the text mining competition (see Section 11.4.2). A comparison of graphs for the preliminary testing and the contest entry show similar performance for a majority of the 22 anomaly classes. Some of the best curves of the true-positive rate (TPR) versus the false-positive rate (FPR) were obtained for the following anomalies: 4, 5, 7, 11, 14, 12, 21, and 22. The worst predictions were obviously for anomalies 2, 10, and 13. In the cases of anomalies 7, 11, 15, and 18, the performance obtained in the contest entry (as measured by the area under the respective ROC curves) was slightly better than that obtained in the preliminary testing.

Thirteen (of the 22) event types (or anomaly descriptions) listed in Table 11.3 were obtained from the Distributed National ASAP Archive (DNAA) maintained by the University of Texas Human Factors Research Project.[4] The generality of topics described in the ASRS reports of the *Noncompliance* (anomaly 2), *Uncommanded (loss of control)* (anomaly 10), and *Weather Issue* (anomaly 13) categories greatly contributed to the poorer performance of the NMF classifier. Additional experiments with a larger numbers of features ($k > 40$) may produce an NMF model that would better capture the diversity of contexts described by those events.

11.5 Summary and Future Work

Nonnegative matrix factorization (NMF) is a viable alternative for automated document classification problems. As the volume and heterogeneity of documentation continues to grow, the ability to discern common themes and contexts can be problematic. This study demonstrated how NMF can be used to both learn and assign (anomaly) labels for documents from the ASRS. As this study was done as part of a text mining competition and a project in a data mining course taught at Wake Forest University, there is room for improvement in both the performance and interpretability of the NMF. In particular, the summarization of anomalies (document classes) using k NMF features needs further work. Alternatives to the filtering of elements of

---

[4] See http://homepage.psy.utexas.edu/HomePage/Group/HelmreichLAB.

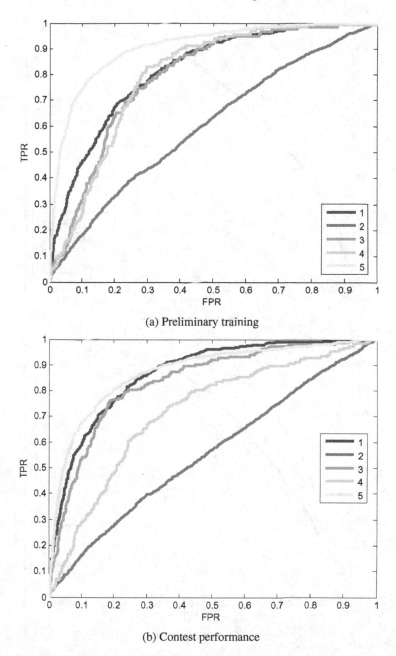

(a) Preliminary training

(b) Contest performance

Fig. 11.1. ROC curves for the NSF classifier applied to anomalies (labels) 1 through 5.

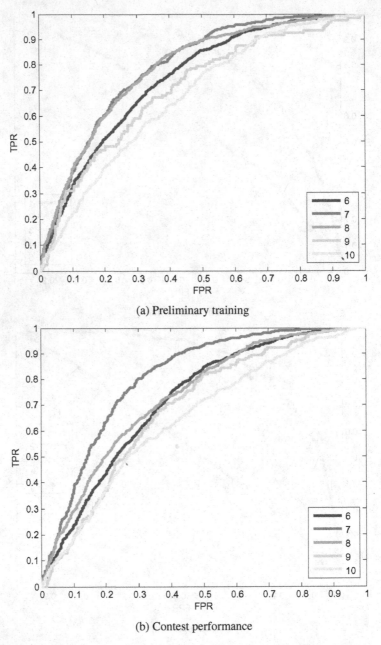

(a) Preliminary training

(b) Contest performance

Fig. 11.2. ROC curves for the NSF classifier applied to anomalies (labels) 6 through 10.

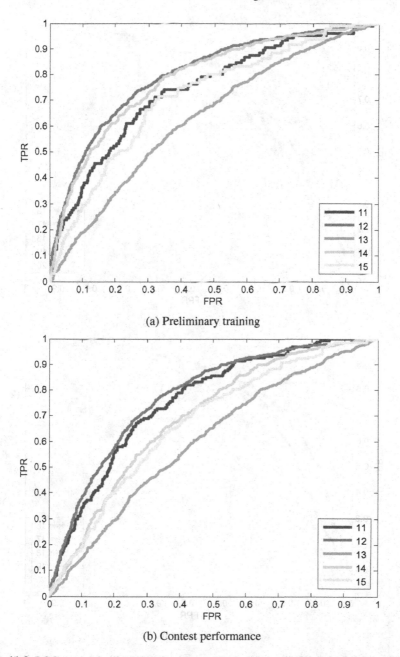

(a) Preliminary training

(b) Contest performance

Fig. 11.3. ROC curves for the NSF classifier applied to anomalies (labels) 11 through 15.

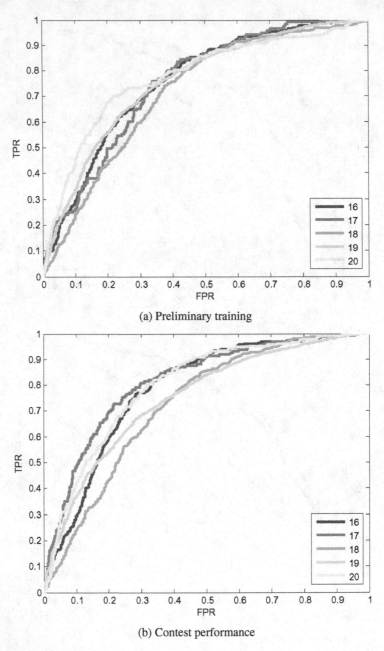

(a) Preliminary training

(b) Contest performance

Fig. 11.4. ROC curves for the NSF classifier applied to anomalies (labels) 16 through 20.

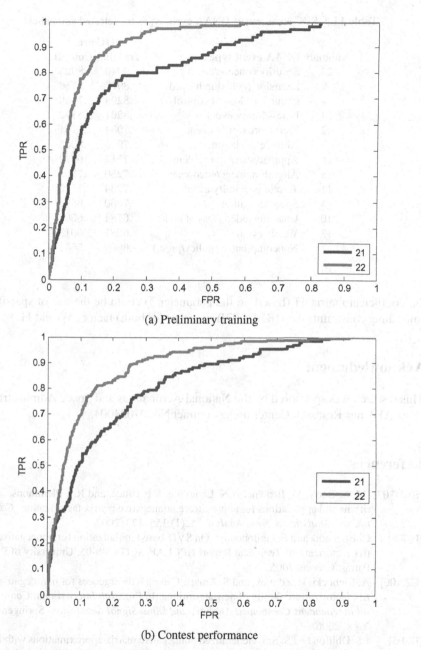

(a) Preliminary training

(b) Contest performance

Fig. 11.5. ROC curves for the NSF classifier applied to anomalies (labels) 21 and 22 in the TRAINCATEGORYMATRIX.CSV file).

Table 11.3. ROC areas versus DNAA event types for selected anomalies

		ROC area	
Anomaly	DNAA event type	Training	Contest
22	Security concern/threat	.9040	.8925
5	Incursion (collision hazard)	.8977	.8716
4	Excursion (loss of control)	.8296	.7159
21	Illness/injury event	.8201	.8172
12	Traffic proximity event	.7954	.7751
7	Altitude deviation	.7931	.8085
15	Approach/arrival problems	.7515	.6724
18	Aircraft damage/encounter	.7250	.7261
11	Terrain proximity event	.7234	.7575
9	Speed deviation	.7060	.6893
10	Uncommanded (loss of control)	.6784	.6504
13	Weather issue	.6287	.6018
2	Noncompliance (policy/proc.)	.6009	.5551

the coefficient matrix H (based on the parameter δ) could be the use of sparsity or smoothing constraints (see [BBL$^+$07]) on either (or both) factors W and H.

Acknowledgment

This research was sponsored by the National Aeronautics and Space Administration (NASA) Ames Research Center under contract No. 07024004.

References

[BBL$^+$07] M.W. Berry, M. Browne, A.N. Langville, V.P. Pauca, and R.J. Plemmons. Algorithms and applications for approximate nonnegative matrix factorization. *Computational Statistics & Data Analysis*, 52(1):155–173, 2007.

[BG05] C. Boutsidis and E. Gallopoulos. On SVD-based initialization for nonnegative matrix factorization. Technical Report HPCLAB-SCG-6/08-05, University of Patras, Patras, Greece, 2005.

[CZA06] A. Cichocki, R. Zdunek, and S. Amari. Csiszár's divergences for non-negative matrix factorization: family of new algorithms. In *Proc. 6th International Conference on Independent Component Analysis and Blind Signal Separation*, Springer, New York, 2006.

[DS05] I.S. Dhillon and S. Sra. Generalized nonnegative matrix approximations with Bregman divergences. In *Proceeding of the Neural Information Processing Systems (NIPS) Conference*, Vancouver, B.C., 2005.

[GBV01] D. Guillamet, M. Bressan, and J. Vitria. A weighted non-negative matrix factorization for local representations. In *Proc. 2001 IEEE Computer Society Conference on Computer Vision and Pattern Recognition*, volume 1, pages 942–947, IEEE, Los Alamitos, CA, 2001.

[GWB03] J.T. Giles, L. Wo, and M.W. Berry. GTP (General Text Parser) software for text mining. In H. Bozdogan, editor, *Software for Text Mining, in Statistical Data Mining and Knowledge Discovery*, pages 455–471. CRC Press, Boca Raton, FL, 2003.

[GZ05]   E.F. Gonzalez and Y. Zhang. Accelerating the Lee-Seung Algorithm for Nonnegative Matrix Factorization. Technical Report TR-05-02, Rice University, March 2005.

[HB06]   A. B. Hamza and D. Brady. Reconstruction of reflectance spectra using robust non-negative matrix factorization. *IEEE Transactions on Signal Processing*, 54(9):3637–3642, 2006.

[Lin05a] C.-J. Lin. On the Convergence of Multiplicative Update Algorithms for Nonnegative Matrix Factorization. Technical Report Information and Support Services Technical Report, Department of Computer Science, National Taiwan University, 2005.

[Lin05b] C.-J. Lin. Projected gradient methods for non-negative matrix factorization. Technical Report Information and Support Services Technical Report ISSTECH-95-013, Department of Computer Science, National Taiwan University, 2005.

[LS99]   D. Lee and H. Seung. Learning the parts of objects by non-negative matrix factorization. *Nature*, 401:788–791, 1999.

[LS01]   D. Lee and H. Seung. Algorithms for non-negative matrix factorization. *Advances in Neural Information Processing Systems*, 13:556–562, 2001.

[Paa97] P. Paatero. Least squares formulation of robust non-negative factor analysis. *Chemometrics and Intelligent Laboratory Systems*, 37:23–35, 1997.

[Paa99] P. Paatero. The multilinear engine—a table-driven least squares program for solving multilinear problems, including the n-way parallel factor analysis model. *Journal of Computational and Graphical Statistics*, 8(4):1–35, 1999.

[PPA07] V.P. Pauca, R.J. Plemmons, and K. Abercromby. Nonnegative Matrix Factorization Methods with Physical Constraints for Spectral Unmixing, 2007. In preparation.

[PT94]   P. Paatero and U. Tapper. Positive matrix factorization: a non-negative factor model with optimal utilization of error estimates of data values. *Environmetrics*, 5:111–126, 1994.

[WCD03] S. Wild, J. Curry, and A. Dougherty. Motivating non-negative matrix factorizations. In *Proceedings of the Eighth SIAM Conference on Applied Linear Algebra*, SIAM, Philadelphia, 2003. Available from World Wide Web: http://www.siam.org/meetings/la03/proceedings.

[WJHT04] Y. Wang, Y. Jiar, C. Hu, and M. Turk. Fisher non-negative matrix factorization for learning local features. In *Asian Conference on Computer Vision*, Korea, January 27–30, 2004.

[ZC06]   R. Zdunek and A. Cichocki. Non-negative matrix factorization with quasi-newton optimization. In *Proc. Eighth International Conference on Artificial Intelligence and Soft Computing, ICAISC*, Zakopane, Poland, June 25–29, 2006.

12

Document Representation and Quality of Text: An Analysis

Mostafa Keikha, Narjes Sharif Razavian, Farhad Oroumchian, and
Hassan Seyed Razi

Overview

There are three factors involved in text classification: the classification model, the similarity measure, and the document representation. In this chapter, we will focus on document representation and demonstrate that the choice of document representation has a profound impact on the quality of the classification. We will also show that the text quality affects the choice of document representation. In our experiments we have used the centroid-based classification, which is a simple and robust text classification scheme. We will compare four different types of document representation: N-grams, single terms, phrases, and a logic-based document representation called RDR. The N-gram representation is a string-based representation with no linguistic processing. The single-term approach is based on words with minimum linguistic processing. The phrase approach is based on linguistically formed phrases and single words. The RDR is based on linguistic processing and representing documents as a set of logical predicates. Our experiments on many text collections yielded similar results. Here, we base our arguments on experiments conducted on Reuters-21578 and contest (ASRS) collection (see Appendix). We show that RDR, the more complex representation, produces more effective classification on Reuters-21578, followed by the phrase approach. However, on the ASRS collection, which contains many syntactic errors (noise), the 5-gram approach outperforms all other methods by 13%. That is because the 5-gram approach is a robust method in presence of noise. The more complex models produce better classification results, but since they are dependent on natural language processing (NLP) techniques, they are vulnerable to noise.

12.1 Introduction

Text classification is the task of assigning one or more classes to a passage from a predefined set of classes, and it can be done in different ways. The distribution of the class members among the training and test dataset is an important factor in determining the success of the classification algorithm.

Most of the classification algorithms work based on a defined distance between two documents, and this distance can also be computed based on different features. The choice of the feature set has a profound effect on the quality of the classification. Different document representation methods create different types of features.

There are many different document representations. The simplest is the N-gram where words are represented as strings of length N. The most popular and effective representation is single words, where documents are represented by their words. Most often the stem of the words is used instead of the words themselves. Stemming the words increases the possibility of matches between the documents and the queries or other documents. A little more sophisticated approach involves extracting statistical or linguistic phrases and representing documents with their stemmed single words and phrases. All the above approaches assume term independence, that is, the relevance of one term does not provide any clues for relevance of other terms.

There are document representation models that do not assume term independence and are able to represent term relationships. One such approach is logical imaging [CvR95]. Other systems such as DRLINK have complex document representation where documents are represented by several features that were extracted and represented independent of each other. For example, one of many features of the DRLINK system [LPY94] is called subject field coder (SFC). The SFCs were codes assigned to topics. The documents were processed using NLP techniques, and then several SFC codes were assigned to each document. SFCs also were matched against SFCs in other documents and queries and the result was combined with the results of other document features to produce the final similarity weight. Some systems use logic to represent the relationships among the words in the text. One such system is PLIR [OO96]. PLIR uses NLP-based methods to discover the relationship among different words or phrases in the collection. Then it uses a logical form to represent a document. This logical representation is called rich document representation (RDR). In this chapter we want to show that in the context of document classification, the quality of the representation is of utmost importance. Also we want to demonstrate that degrading the quality of text affects the finer grain representations more than the simple and crude representations.

In Section 12.2 we will describe the vector space model with more detail. Section 12.3 describes our classification approach, and in Section 12.4 the document representation approaches will be discussed along with the distance and similarity measures. Section 12.5 describes the experimental results, and Section 12.6 presents the conclusion of the obtained results.

12.2 Vector Space Model

The vector space is one of the most common models for representing documents. In this model, each document is represented as a vector of *terms*. The terms are the features that best characterizes the document and can be anything from a string of length N, single words, phrases, or any set of concepts or logical predicates. The VSM does not keep any information regarding the order in which the terms occur. Often, before

processing the terms, stop-words, terms with little discriminatory power, are eliminated. Also it is common to use the stems of the words instead of the actual words
themselves.

All the terms in the dataset define a "space" in which each term represents one
"dimension." For distinguishing a document from the other documents, numeric values are assigned to terms in order to show the importance of that term for that document.

The base weighting schema in vector space model uses two main factors: the
frequency of a term in the document or term frequency (tf), and the inverse of the
number of documents containing that term or inverse document frequency (idf). The
tf factor indicates the importance of a term in the document and is a document-
specific statistic. This parameter is usually normalized. The idf factor is a global
statistic and measures how widely a term is distributed over the collection. There are
many functions for calculating tf and idf factors described in [SB88].

Another element in weighting schema is the normalization factor, which usually tries to diminish the effect of the length of the document in weighting. We will
explain our normalization factor in the weighting subsection [Gre00].

12.3 Text Classification

There is a wide range of classification methods available for the task of text classification. The support vector machines (SVM) method [YL99] is based on a learning
approach to solve the two-class pattern recognition problems. The method is defined
over a vector space where the problem is to find a decision surface that best separates the documents into two classes. The k-nearest neighbor (kNN) method, on the
other hand, classifies documents according to the voting of its k nearest neighbor
documents, which can be discovered using a distance measure between documents
[YL99]. The neural network has also been used in classification tasks. However, Han
and Karypis showed that the two methods of SVM and kNN significantly outperform
the neural network approach when the number of the positive training instances per
category is small [HK00]. Another method is the centroid-based text classification
[HK00], which is a very simple yet powerful method that outperforms other methods
on a wide range of datasets. In this method, given a new document, D, which is to be
labeled, the system computes the similarity of D to the centroid of each of the existing classes. The centroid of a class is defined as the average of the document vectors
of that class. The following formula shows how the centroid of a class is calculated:

$$C_k = \frac{\sum_{D_i \in Class_k} \vec{D_i}}{|Class_k|} \tag{12.1}$$

where C_k is the kth centroid, D_i is the vector representation of the ith document,
and $Class_k$ is the collection of documents that are members of the kth class. Any
document that is to be classified is compared with all of the centroids of the clusters,
and receives a similarity value for each of the centroids:

$$Sim(\overrightarrow{D_{new}}, \overrightarrow{C_i}) = \overrightarrow{D_{new}}.\overrightarrow{C_i} \qquad (12.2)$$

After comparing with all of the centroids, the system computes a score of membership for that document in each one of the classes. These scores form the category vector for that document

A threshold could then be used in calculation of the final categories. It is noticeable that working with category vectors enables the method to deal with multiple categories for a document.

12.4 Document Representation Approaches

Document representation is the method of representing documents in a system. There are many models for representing documents, which differ on the assumptions they make about the words and documents. The two common assumptions are term independence and document independence.

Term independence states that from the relevance of a term we cannot make any statement about the relevance of other terms. Term independence assumes that terms in documents of a collection have no relationship with each other. Document independence states that the relevance of a document has no effect on the relevance of the other documents. The validity of these assumptions or lack of it has nothing to do with their usefulness. The simpler document representation models, by accepting both of the above assumptions, treat documents as a bag of terms or a set of independent features. The more complex models may accept only one or none of the above assumptions. For example, systems that use a thesaurus accept only limited term independence where the only accepted relationships between terms are thesaurus relations such as similarity or dissimilarity. IR systems that use clustering in grouping their output do not subscribe to document independence; therefore, they try to group the like documents in to one group. Logic-based systems such as GRANT [KC88] do not subscribe to neither of the assumptions. The GRANT system follows a semantic network approach where all the terms and documents are connected to each other through a variety of relationships.

In this chapter we investigate two questions. First, is there a relationship between the complexity of the representation and the quality of the classification? Second, is there a relationship between the quality of text and the performance of the document representations? To answer these questions we will look at four different document representations with different levels of complexity.

However, the way a system uses its document representation is also an issue. That is, two systems with identical document representations but with different matching techniques will perform differently. For example, RDR is a logical representation that does not subscribe to either of the independence assumptions. The PLIR system utilizes RDR as a document representation model. PLIR uses reasoning in order to estimate the similarity of queries to documents or documents to each other. The power of PLIR, which normally outperforms the best vector space models in experimental conditions, comes from both its document representation and its inferences.

To isolate the effect of document representation, we will apply the same weighting and matching techniques to different document representation models. For this purpose we have selected the VSM because of its simplicity. We will demonstrate the performance of four document representation models—N-grams, single words, single words plus phrases, and RDR—in the presence and absence of noise to show how noise affects quality and what can be done.

12.4.1 N-Grams

In the N-gram representation approach, the text is broken down into strings of n consecutive characters with or without regard to word length or word boundaries [Gre00].

Zamora uses trigram analysis for spelling error detection [ZPZ81]. Damashek uses N-grams of length 5 and 6 for clustering of text by language and topic. He uses $N = 5$ for English and $N = 6$ for Japanese [Dam95]. Some authors [Sue79] draw N-grams from all the words in a document but use only N-grams wholly within a single word. Others [Dam95] use N-grams that cross word boundaries; that is, an N-gram string could start within one word and end in another word, and include the space characters that separate consecutive words.

A pure N-gram analysis does not use language-specific or semantic information. Stemming, stop-word removal, syntactically based phrase detection, thesaurus expansion, etc., are ignored by this method. So, theoretically, the performance of this approach should be lower than methods that make effective use of the language-specific clues. However, it is a very simple and fast method, and can be a very effective in situations where the language of the document is not previously known, or when the dataset contains textual errors. In some languages such as Persian, N-gram methods have comparable performance to that of unstemmed single words [AHO07]. Authors of [Dam95] and [PN96] show that the performance of an N-gram system is remarkably resistant to textual errors, for example, spelling errors, typos, errors associated with optical character recognition, etc. In our experiments, we used 3, 4, and 5 as N and didn't cross the word boundaries in the creation of the N-grams. We used words without stemming and did not remove stop words. When the n in the N-gram method increases, the method slowly loses its N-gram characteristic, and gains more of the characteristics of the single word representation method. For a special dataset such as the SIAM text mining competition dataset, we noticed that the 5-gram approach results in the best performance as explained below.

12.4.2 Single-word representation

Single words are the most common way by which to represent documents. In this method, each document is represented as a vector of weights of its distinct words.

This method, while quite simple, is very powerful for indexing documents [Lee94]. However, some characteristics of the documents may affect the performance of this method. Spelling errors, for instance, cause the incorrect weights to be assigned to words. The vector space model and the $TFIDF$ weighting are very sensitive to weights. In these situations, errors will result in false-weights [BDO94]. Preprocessing algorithms like error detection and correction can be helpful in these situations.

12.4.3 Stemmed Single-Word Representation

Stemming is a method to improve the quality of single-word indexing by grouping words that have the same stem. The most common stemming algorithm for English language is the Porter algorithm [Por80]. This algorithm removes the common post-fixes in different steps, and its simplicity and high effectiveness have caused it to be used in many applications requiring stemming. In our experiments we used this algorithm for stemming.

12.4.4 Phrases

Many systems index phrases along with single words. There are two ways to form phrases. First is statistical, where co-occurrence information is used in some way to group together words that co-occur more than usual. Second, is syntactical, where linguistic information is used to form the phrases. For example, an adjective and a noun together form a phrase. Normally the length of statistical phrases is two. But in systems that use linguistic phrases the length of the phrases is one the system parameters. It could be two, three, or more. There is little value in using statistical phrases. However, syntactical phrases are more widely used in conjunction with stemmed words in order to improve the precision of the system.

12.4.5 Rich Document Representation (RDR)

Another approach in document indexing is rich document representation. In this method, the document is represented by a set of logical terms and statements. These logical terms and statements describe the relationships that have been found in the text with a logical notation close to the multivalued logic of Michalski. In [JO04] it has been reported that this kind of document indexing has improved the quality of the document clustering.

The process of producing these logical forms is as follows: First, the text is tagged by a part-of-speech (POS) tagger. Then a rule-based extraction process is applied to the output of the part of speech tagger. Matched rules indicate the existence of a special relation in the text. For example, a proposition such as "for" in the sentence fragment such as "...*operating systems for personal computers*..." suggests a relationship between two noun phrases "*operating systems*" and "*personal computers*" [RKOR06]. Then, these relations are represented with a format similar to that of multivalued logic as used in the theory of human plausible reasoning; that is, *operating_system(personal_computers)* [CM89]. In a similar way, a logical statement such as *operating_system (personal_computers) = Windows_XP* can be formed from a sentence fragment such as "... *Windows XP is a new operating system for personal computers*...."

Rich document representation represents documents by their stemmed single terms, stemmed phrases, logical terms, and logical statements. This method provides more semantic representation for a document. PLIR system uses the RDR and combines all the documents' representations into a single semantic network. By doing this, the information contained in documents complements other information and creates a semantic space. PLIR applies its inferences in this semantic space to infer relevance or closeness of documents to concepts or other documents [OO96]. The performance of this method depends on the power of the rules and the characteristics of the dataset. Noisy text usually misleads the part of speech tagger and the pattern matching rules, and thus reduces the performance of the method. Also, the ambiguities of the natural language text, for example, the existence of anaphora, make this method susceptible to missing some of the relations. A higher level of text preprocessing in the discourse or pragmatic level can lead to a better performance of this representation model.

12.5 Experiments

Two set of experiments have been performed. In the first set, we used the Reuters-21578 test collection and demonstrated that RDR has better results than N-grams, single words, and single words plus phrases. In the second set of experiments, we used the contest text collection (ASRS documents), and ran our tests with 3-gram, 4-gram, and 5-gram, and single word, stemmed single word, and RDR document representations, and found out that among these experiments 5-gram was better than the other representation methods. The main difference between the first and second set of experiments is the presence of noise in the SIAM collection. In all experiments we used centroid-based text classification as our classification method.

12.5.1 Experiment 1: Reuters-21578 Collection

In the first part of the experiments we used the Reuters-21578 collection. To divide the collection into training and test sets, we used the ModLewis split. This leads to a training set consisting of 13,625 stories and a test set consisting of 6188 stories and 135 different categories.

We created four different representations for each document including 5-gram, single term, and phrase and logical representation. Using these representations, four different similarities for each document and category were computed. Here we specified one category for each document, so the category with the highest similarity will be assigned as the category of that document.

All of the experiments in this part have been done on stemmed words, based on the Porter stemming algorithm. Stop words were also removed using Van Rijsbergen's list for removing stop words [Rij79]. In the weighting step, we used the same weighting schema for both training and test documents, namely $Ltu.Ltu$ weighting schema [Gre00], which means we used the following formulas for tf, idf, and normalization factor (nf):

$$tf = \frac{1 + \log(term freq)}{1 + \log(average\ term freq)} \tag{12.3}$$

$$idf = \ln \frac{N}{n} \tag{12.4}$$

$$nf = \frac{1}{(slope * \#\ of\ unique\ terms) + (1 - slope) * pivot} \tag{12.5}$$

$$pivot = \frac{(\#\ of\ doc\ unique\ terms)}{avg(\#\ of\ doc\ unique\ terms)} \tag{12.6}$$

where:
$term freq$ is the frequency of term in the document;
N is the number of all documents in the collection;
n is the number of documents that contain that term;
$\#\ of\ doc\ unique\ terms$ is the number of unique terms in the specified document;
$avg(\#\ of\ doc\ unique\ terms)$ is the average number of unique terms in all documents.

Multiplication of these three parameters for each term is the weight of that term in the specified document. We did our experiments with two values for slope (0.25 and 0.75). Table 12.1 shows the accuracy for each indexing method. The accuracy is defined as

$$accuracy = \frac{TP}{TP + FP} \tag{12.7}$$

where:
TP is True Positive, which is the number of cases where the correct classification has been identified;
FP is False Positive, which is the number of cases where the category has been incorrectly identified.

Table 12.1 shows the performance of each method in its simplest form. Each run uses only one type of token in its indexes. For example, $Run1$ reports the performance of stemmed single words, $Run3$ uses only stemmed phrases, and $Run5$ contains only stemmed logical terms. Based on this table, the single-term indexing method has the best accuracy.

Of course, more complex document representations include the simpler forms also. Table 12.2 depicts the real performance of more complex models. $Run9$ shows the performance of phrases when the document representation contains stemmed single words and phrases. $Run10$ shows the performance of the RDR method, which contains stemmed single words, stemmed phrases, and logical terms. For computing the weights for Single+Phrase and RDR, the weights of the simpler forms were summed to produce the final similarity. The category assignment stage was repeated for these models with these new similarities. Because the slope of 0.25 produced the best results in Table 12.1 in the combination stage, only the similarity values calculated with the slope of 0.25 were used for summation.

Table 12.1. Accuracy of three different indexing methods

Run.ID	Indexing method	Slope	Accuracy
1	Single	0.25	56.3
2	Single	0.75	56.0
3	Phrase	0.25	53.9
4	Phrase	0.75	53.9
5	Logic	0.25	30.9
6	Logic	0.75	30.9
7	5-Gram	0.25	47.8
8	5-Gram	0.75	47.8

Table 12.2. Accuracy of combined methods

Run.ID	Combined methods	Combined run.IDs	Accuracy
9	Single+Phrase	(1),(2)	59.9
10	RDR	(1),(2),(3)	63.2

Table 12.1 demonstrates that the accuracy of the RDR method is 6.9% better than the accuracy of the single-term indexing and 15.4% better than 5-gram and about 3.3% better than phrases. It should be noted that when more complex combinatoric formulas are used for combining the simple similarity values, the difference between the Single+Phrase model and RDR could be as high as 7% or 9%.

This experiment demonstrates that the more complex a document representation, the better the performance of the categorization. In this experiment, single words outperforms 5-grams, Single+Phrase outperforms single words, and the RDR is better than phrases.

12.5.2 Experiment 2: SIAM Text Mining Contest Collection

In this part of the experiment we used the contest (ASRS) test collection. We indexed the first 20,000 documents of training file with six different indexing schemes discussed in Section 12.3; 3-gram, 4-gram, and 5-gram, and single-word, stemmed-single-word, and RDR.

For indexing each document, the frequency of each term in that document was counted as the term frequency, or $termfreq$ of that term, and the average of $termfreq$ for all document terms was calculated. The final weight of a term in a document was then calculated as

$$\frac{1 + \log(termfreq)}{1 + \log(average\ termfreq)} \tag{12.8}$$

After indexing all the documents, we also calculated the inverse document frequency (idf) in a separate table for each of the terms. Using idf can compensate for the lack of stop-word removal, because stop-words receive a very low idf automatically. We then updated our weight column and multiplied the idf factor to the existing tf factor.

With the arrival of each new document, the same indexing was used to create the new document vector. Each document vector with the above weighting was then multiplied to the centroid vector of each class. To normalize the similarity measure, cosine normalization was used. In cosine normalization, we divided the result of the dot-product of two vectors by their lengths. This way, we are actually computing the cosine of the angle between two vectors.

We used 1519 documents of the training data to test our classification. Based on the real categories of the documents for each of the indexing methods, we calculated the best thresholds that maximized the fraction of correct labels to incorrect labels.

Because of the differences in the distribution of the class members, we calculated the best thresholds for each of the 22 categories separately. For doing so, first the Max_i, the maximum of the scores of all test documents in the ith class was computed. Then we increased the threshold from 0 to Max_i in 10,000 steps, and the calculated true-positive (TP), true-negative (TN), false-negative (FN), and false-positive (FP) results based on that threshold. The best threshold was the one that maximized the following formula:

$$\frac{TP + TN}{FN + FP}. \tag{12.9}$$

This helped the system to increase correct answers and decrease incorrect class assignments. Our initial test created the best thresholds, and we used those thresholds to create the final categories of the contest test data.

Table 12.3 shows the accuracy calculated for each one of document representations. All the runs used the same weighting scheme and thresholds, and the only difference is the document representation. The result obtained in this collection is contrary to the result obtained from the Reuters-21578 collection. In the ASRS collection the more complex models are the least effective and the simplest ones are the most effective ones. In this collection, the 4-gram and 5-gram models have the best performance. The single-term models (with and without stemming) are behind by almost 13 points. The RDR model, which is the most complex among them, is 15 points behind and even worse than single terms without stemming. The main difference between these two sets of experiments is the presence of noise. The ASRS collection has many spelling errors, and that is why the models that are prone to spelling errors are not affected. On the other hand, the models that are heavily dependent on POS tagging, information extraction, and other NLP methods for processing text and creating representations are very vulnerable to textual variations. The NASA-sponsored text mining contest uses the receiver operating characteristic (ROC) curves for measuring the quality of the classification. These curves are two-dimensional measures of classification performance, and show how the number of correctly classified positive examples varies with the number of incorrectly classified negative examples. The authors in [DG06] show that for a classification method, the ROC curve and the classic precision/recall curve have a deep connection, and when

Table 12.3. Accuracy of each method for best threshold value

Run.ID	Indexing method	Accuracy
11	Single	52.3
12	Stemmed Single	53.9
13	Single+Phrase	53.3
14	RDR	51.4
15	5-Gram	66.7
16	4-Gram	65.6
17	3-Gram	59.5

a classifier dominates in the ROC curve, it will dominate in the precision/recall curve too, and vice versa.

In ROC space, one plots the false-positive rate (FPR) on the x-axis and the true-positive rate (TPR) on the y-axis. The FPR measures the fraction of the negative examples that are misclassified as positive. The TPR measures the fraction of positive examples that are correctly labeled. Based on this observation, we compared our categorization results against the real values, and drew the ROC curve for all of them, which can be seen in Figures 12.1 and 12.2.

Here also it is evident that the 5-gram model is the one that dominates the ROC curve. The area under the ROC curve is a standard measure of performance for classifier algorithms [DG06], and that means that the 5-gram model is the best document representation method for this noisy dataset.

12.6 Conclusion

There are three factors in text categorization: categorization model, similarity measure, and document representation. There are many alternatives for each one of these factors. In this chapter we focused on the document representation model and examined four different document representation techniques with different levels of complexity. The simplest of all is the N-gram model where the words are broken down into overlapping strings of length N. The most popular model is the single-term approach where each word is treated as a term independent of any other term. It is common to use word stems instead of their original surface form in order to increase recall. A little more complicated model uses phrases (mostly syntactic) along with single terms. There are many more complex approaches to document representation. Here we used a logical approach called rich document representation (RDR), which extracts the relationships between terms and represents them as predicates. We have demonstrated that in the context of the classification, the better document representation produces better classification. This is not completely the same as ad-hoc information retrieval. Over the years it has been observed that many different systems with different document representations exhibit similar performances in the context of TREC conferences.

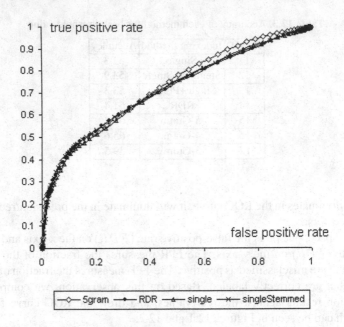

Fig. 12.1. ROC curve for 5-gram, single-word, stemmed single-word, and RDR indexing

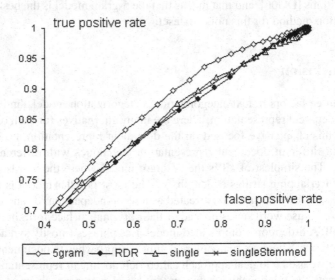

Fig. 12.2. Close-up of ROC curve of Figure 12.1

We have also demonstrated that the noise, the textual variations, has a grave effect on more complex document representations. We showed that on the SIAM dataset, which contains many spelling errors, the best models are the simplest and more robust ones.

In that context, the 5-gram approach dominated the other methods. The worst approach was the RDR model, which is dependent on NLP techniques to extract and represent the relationships. In the text mining contest (ASRS) dataset, our centroid classification method with 5-gram document representation model shows significant improvement over the Schapire's and Singer's BoosTexter approach as standard approaches.

In the context of document classification, the choice of document representation should be taken seriously. It should be decided based on two factors: first, the amount of the noise or syntactic variations in the text collection; second, the amount of time and resources available. The simpler models require less processing, while the more complicated models such as RDR require POS tagging and other NLP text processing techniques. The NLP methods and tagging are time-consuming and heavy users of resources.

Acknowledgments

This work is partially supported by the Iranian Communication Research Center (ITRC) under contract number 500/12204.

References

[AHO07] A. AleAhmad, P. Hakimian, and F. Oroumchian. N-gram and local context analysis for persian text retrieval. *International Symposium on Signal Processing and its Applications (ISSPA2007)*, 2007.

[BDO94] M.W. Berry, S.T. Dumais, and G.W. O'Brien. Using linear algebra for intelligent information retrieval. Technical Report UT-CS-94-270, University of Tennessee, 1994. Available from World Wide Web: citeseer.ist.psu.edu/berry95using.html.

[CM89] A. Collins and R. Michalski. The logic of plausible reasoning: a core theory. *Cognitive Science*, 13(1):1–49, 1989. Available from World Wide Web: citeseer.ist.psu.edu/collins89logic.html.

[CvR95] F. Crestani and C.J. van Rijsbergen. Probability kinematics in information retrieval. In *Proceedings of the Eighteenth Annual International ACM SIGIR Conference on Research and Development in Information Retrieval*, pages 291–299, ACM Press, New York, 1995.

[Dam95] M. Damashek. Gauging similarity with n-grams: language-independent categorization of text. *Science*, 267(5199):843, 1995.

[DG06] J. Davis and M. Goodrich. The relationship between Precision-Recall and ROC curves. *Proceedings of the 23rd International Conference on Machine Learning*, pages 233–240, ACM Press, New York, 2006.

[Gre00]   E. Greengrass. Information Retrieval: A Survey. *IR Report*, 120600, 2000. Available from World Wide Web: http://www.csee.umbc.edu/cadip/ readings/IR.report.120600.book.pdf.

[HK00]   E.H. Han and G. Karypis. *Centroid-based Document Classification: Analysis and Experimental Results*. Springer, New York, 2000.

[JO04]   A. Jalali and F. Oroumchian. Rich document representation for document clustering. In *Coupling Approaches, Coupling Media and Coupling Languages for Information Retrieval Avignon (Vaucluse)*, pages 800–808, RIAO, Paris, France, 2004.

[KC88]   R. Kjeldsen and P.R. Cohen. The evolution and performance of the GRANT system. *IEEE Expert*, pages 73–79, 1988.

[Lee94]   J.H. Lee. Properties of extended Boolean models in information retrieval. *Proceedings of the 17th Annual International ACM SIGIR Conference on Research and Development in Information Retrieval*, pages 182–190, ACM Press, New York, 1994.

[LPY94]   E.D. Liddy, W. Paik, and E.S. Yu. Text categorization for multiple users based on semantic features from a machine-readable dictionary. *ACM Transactions on Information Systems (TOIS)*, 12(3):278–295, 1994.

[OO96]   F. Oroumchian and R.N. Oddy. An application of plausible reasoning to information retrieval. In *Proceedings of the Nineteenth Annual International ACM SIGIR Conference on Research and Developement in Information Retrieval*, pages 244–252, ACM Press, New York, 1996.

[PN96]   C. Pearce and C. Nicholas. TELLTALE: Experiments in a dynamic hypertext environment for degraded and multilingual data. *Journal of the American Society for Information Science*, 47(4):263–275, 1996.

[Por80]   M.F. Porter. An algorithm for suffix stripping. *Information Systems*, 40(3):211–218, 1980.

[Rij79]   C.J. Van Rijsbergen. *Information Retrieval*. Butterworth-Heinemann, Newton, MA, 1979.

[RKOR06] F. Raja, M. Keikha, F. Oroumchian, and M. Rahgozar. Using Rich Document Representation in XML Information Retrieval. *Proceedings of the Fifth International Workshop of the Initiative for the Evaluation of XML Retrieval (INEX)*, Springer, New York, 2006.

[SB88]   G. Salton and C. Buckley. Term weighting approaches in automatic text retrieval. *Information Processing and Management*, 24(5):513–523, 1988.

[Sue79]   C.Y. Suen. N-gram statistics for natural language understanding and text processing. *IEEE Transactions on Pattern Analysis and Machine Intelligence*, 1(2):164–172, 1979.

[YL99]   Y. Yang and X. Liu. A re-examination of text categorization methods. *Proceedings of the 22nd Annual International ACM SIGIR Conference on Research and Development in Information Retrieval*, pages 42–49, ACM Press, New York, 1999.

[ZPZ81]   E.M. Zamora, J.J. Pollock, and A. Zamora. The use of trigram analysis for spelling error detection. *Information Processing and Management*, 17(6):305–316, 1981.

A

Appendix: SIAM Text Mining Competition 2007

Overview

The 2007 Text Mining Workshop held in conjunction with the Seventh SIAM International Conference on Data Mining was the first to feature a text mining competition. Members of the Intelligent Data Understanding group at NASA Ames Research Center in Moffett Field, California, organized and judged the competition. Being the first such competition held as a part of the workshop, we did not expect the large number of contestants that more established competitions such as KDDCUP[1] have, but we did receive five submissions, though one person later withdrew.

Matthew E. Otey, Ashok N. Srivastava
Santanu Das, and Pat Castle

A.1 Classification Task

The contest focused on developing text mining algorithms for document classification. The documents making up the corpus used in the competition are aviation safety reports documenting one or more problems that occurred on certain flights. These documents come from the Aviation Safety Reporting System (ASRS), and they are publicly available in their raw form at http://asrs.arc.nasa.gov. Since each of these documents can describe one or more anomalies that occurred during a given flight, they can be tagged with one or more labels describing a class of anomalies. The goal is to label the documents according to the classes to which they belong, while maximizing both precision and recall, as well as the classifier's confidence in its labeling. This second criterion concerning confidence is useful for presenting results to end-users, as an end-user may be more forgiving of a misclassification if he or she knew that the classifier had little confidence in its labeling.

---

[1] See http://www.kdnuggets.com/datasets/kddcup.html.

The competition entries were scored using the following cost function that accounts for both the accuracy and confidence of the classifier. When describing the cost function, we will use the following notation. Let D be the number of documents in the corpus, and let L be the number of labels. Let F_j be the fraction of documents having label j. Let $t_{ij} \in \{-1, +1\}$ be the target (true) value for label j of document i. Let $p_{ij} \in \{-1, +1\}$ be the predicted value for label j of document i, and let $q_{ij} \in [0, 1]$ be the classifier's confidence of the value p_{ij}. Finally, let A_j be the area under the ROC curve for the classifier's predictions for label j. We define the intermediate cost function Q_j for label j as

$$Q_j = (2A_j - 1) + \frac{1}{D} \sum_{i=1}^{D} q_{ij} t_{ij} p_{ij} \, . \tag{A.1}$$

This function has a maximum value of 2. The final cost function Q is the average Q_j for all labels:

$$Q = \frac{1}{C} \sum_{j=1}^{C} Q_j \, . \tag{A.2}$$

In case of ties, we also formulated a figure of merit (FOM), defined as

$$FOM = \frac{1}{C} \sum_{j=1}^{C} \frac{(F - F_j)}{F} Q_j \tag{A.3}$$

where

$$F = \sum_{j=1}^{C} F_j. \tag{A.4}$$

This figure of merit measures the quality of predictions across all anomaly categories, giving a lower weighting to those categories making up a larger fraction of the dataset. Hence, better FOM are achieved by obtaining higher accuracies and confidences on rarer classes.

We wrote a small Java program implementing these functions to use during our judging of the submissions. Before the contest deadline, we made this program and its source code available to the contestants so that they could validate its correctness, and so that they could use our implementation to tune their algorithms.

A.2 Judging the Submissions

A training dataset was provided over a month in advance of the deadline, giving the contestants time to develop their approaches. Two days before the deadline we released the test dataset. Both of these datasets had been run through PLADS, a system that performs basic text processing operations such as stemming and acronym expansion. The contestants submitted their labeling of the test dataset, their confidences in the labeling, and the source code implementing their approach. The scores of the submissions were calculated using the score function described above. In addition to scoring the submissions, we ran each contestant's code to ensure that it worked and produced the same output that was submitted, and we inspected the source code to ensure that the contestants properly followed the rules of the contest.

A.3 Contest Results

The submissions of the contestants all successfully ran and passed our inspection, and we announced our three winners. First place was awarded to Cyril Goutte at the NRC Institute for Information Technology, Canada, with a score of 1.69. A team consisting of Edward G. Allan, Michael R. Horvath, Christopher V. Kopek, Brian T. Lamb, and Thomas S. Whaples of Wake Forest University, and their advisor, Michael W. Berry of the University of Tennessee, Knoxville, came in second with a score of 1.27. The third place team of Mostafa Keikha and Narjes Sharif Razavian of the University of Tehran in Iran, and their advisor, Farhad Oroumchian of the University of Wollongong, Dubai, United Arab Emirates, scored a 0.97. At NASA, we evaluated Schapire's and Singer's BoosTexter approach, and achieved a maximum score of 0.82 on the test data, showing that the contestants made some significant improvements over standard approaches.

Index